AMERICAN MEDICAL ASSOCIATION

NUTRIENTS IN PROCESSED FOODS

vitamins · minerals

PUBLISHING SCIENCES GROUP, INC.
Acton, Massachusetts
a subsidiary of CHC Corporation

Printed in the United States of America.

International Standard Book Number: 0-88416-006-8

Library of Congress Catalog Card Number: 73-85400

Howard E. Bauman, PhD
Vice President for Science
and Technology
Pillsbury Co.
Minneapolis, Minn.

Clinton L. Brooke
Consultant
United States Agency for
International Development
Raleigh, N.C.

Gerald F. Combs, PhD
Professor and Head
Department of Foods and Nutrition
School of Home Economics
University of Georgia
Athens, Ga.

Philip H. Derse, MS
President of the WARF Institute, Inc.
Madison, Wis.

Richard P. Farrow
Assistant Director of the
Washington Research Laboratory
of the National Canners Association
Washington, D.C.

Lloyd J. Filer, Jr., MD, PhD
Professor in the Department
of Pediatrics
University of Iowa Hospitals
and Clinics
Iowa City, Iowa

R. Gaurth Hansen, PhD
Professor of
Biochemistry and Nutrition
Utah State University
Logan, Utah

Richard E. Hein, PhD
General Manager of
Research and Development
Heinz
U.S.A.
Division of H.J. Heinz Co.
Pittsburgh, Pa.

Imri J. Hutchings, PhD
Research Consultant
H.J. Heinz Co.
Pittsburgh, Pa.

Ogden C. Johnson, PhD
Office of Nutrition and
Consumer Sciences
Food and Drug Administration
Department of Health,
Education, and Welfare
Washington, D.C.

Oral Lee Kline, PhD
Executive Officer
American Institute of Nutrition
Bethesda, Md.

Walter Mertz, MD
Chairman of the
Nutrition Institute
Agricultural Research Service
United States Department
of Agriculture
Beltsville, Md.

James M. Reed
National Canners Association
Washington, D.C.

Herbert P. Sarett, PhD
Vice President
Nutritional Science Resources
Mead Johnson Research Center
Evansville, Ind.

William H. Sebrell, Jr., MD
R.R. Williams Professor Emeritus
of Nutrition
formerly Director
Institute of Human Nutrition
College of Physicians and Surgeons
Columbia University
New York, N.Y.
presently Medical Advisor
Weight Watchers International, Inc.
Great Neck, N.Y.

Frederic R. Senti, PhD
Agricultural Research Service
United States Department
of Agriculture
Washington, D.C.

Ira I. Somers, PhD
Executive Vice President
of the National Canners Association
Washington, D.C.

Frederick J. Stare, MD
Professor of Nutrition
and Chairman
Department of Nutrition
Harvard University School
of Public Health
Boston, Mass.

Philip L. White, ScD
Secretary
American Medical Association's
Council on Foods
and Nutrition
Chicago, Ill.

Virgil O. Wodicka, PhD
Director
Bureau of Foods
Food and Drug Administration
Department of Health,
Education, and Welfare
Washington, D.C.

table of contents

list of illustrations

list of tables

"The maintenance of a high level of nutritional health requires the continuous availability of a wholesome, nourishing food supply that can provide all of the essential nutrients in amounts sufficient to meet human needs." So stated the Council on Foods and Nutrition of the American Medical Association in a policy statement, "Improvement of the Nutritive Quality of Foods." The assurance of a nourishing food supply is not a matter to be taken lightly by the food industry, the government, or the public. When, despite good manufacturing practices, significant nutrient losses occur in food processing, the addition of nutrients through restoration, fortification, or enrichment is recognized as a worthy enterprise. The addition of nutrients to foods, however, is not justifiable as a means of compensating for poor manufacturing practices.

The AMA, through its Council on Foods and Nutrition, has for a good many years published norms and concepts for the nutritional improvement of foods. The approach has been conservative in order to discourage undue artificiality in the food supply. The Council has long maintained that foods are the preferred source of nutrients. The rapid advances by the processed food industry have changed the appearance of markets and pantry shelves. More of the food preparation once done in the home is now performed by the industry, leading to great numbers of convenience foods. With its AMA-Food Industry Liaison Committee, the Council on Foods and Nutrition has been exploring the safety and quality of processed foods.

The exploration has been done through the media of symposia carefully designed to obtain the optimum amount of information pertaining to the influence of food processing technology on the nutritional quality of food. At the same time, human nutrient requirements have been reviewed to permit the drawing of parallels between need and supply. In certain instances, as with food fortification, Federal Regulations have been evaluated as they aid or hinder efforts to assure adequate supplies of nutrients.

The first of the present series of symposia was held in March 1971 and the last in October 1973. This three-volume series, Nutrients in Processed Foods, is a compilation of papers, discussions and recommendations from the symposia. Earlier papers have been brought up to date by the contributing authors.

The first volume examines the effects of processing upon the vitamin and mineral content of foods. This volume is divided into four parts. The first part addresses itself to the vitamin and mineral needs of the general population of the United States, examining evidence of nutrient deficiencies and the effects of excessive intake of micronutrients. Problems encountered in the definition of need for some of the trace nutrients are discussed.

Factors influencing the vitamin and mineral content of foods are treated in Part II. It is here that agriculture, processing, storage and distribution practices are considered as determinants of the nutritive value of foods. Problems encountered in the assay of nutrients and in the evaluation of biological availability are discussed.

All aspects of the nutritional improvement of processed foods, from historical precedence to current regulations, are treated in Part III. Several chapters are devoted to proposals for the classification of food; considerations of patterns of food use for the purpose of encouraging more innovative approaches to food fortification, as need arises; and consumer education on a continuing basis.

Participants at the symposium from which this volume is derived met in groups to develop specific recommendations in five areas of concern: (1) needs for future research; (2) fortification guidelines; (3) education of consumers; (4) regulations; and (5) the rationale and criteria for micronutrient improvement of foods. Part IV contains the reports of the group deliberations which we hope will be given serious consideration by all who read this volume.

The second volume, Proteins, is divided into five parts to permit an orderly discussion of the factors that influence the amount and quality of proteins in foods. The volume is introduced with a review of human protein requirements including requirements under disease conditions requiring special dietary therapy. Part II is concerned with the assessment of protein quality and consideration is given to current methods of evaluating quality— from chemical and small animal assays to direct studies of human requirements.

The third part of this second volume, "Factors Influencing Protein Quality," deals with newer studies of factors that decrease or enhance the availability of amino acids and proteins in foods. It is in this part that discussions of how food processing methods influence protein quality are found. The fourth and fifth parts are devoted to specific task force reports of major concerns in protein nutrition: assessment and application of human requirements, protein quality evaluation, availability of amino acids, and practical approaches to protein improvement. The reader can enjoy reliving the symposium as the task forces digest the material presented and report their opinions and recommendations.

An inordinate amount of concern has centered on fats, sugars and other carbohydrates in foods. The general concerns include fats and carbohydrates as energy sources with profound effects on nutrient density, ie, caloric dilution of the concentration of nutrients; their possible relationship to disease, including heart disease, cancer and dental caries; the role of high concentrations of carbohydrates and fats in determining taste preferences and therefore food selection. The third volume, *Fats and Carbohydrates*, takes up these matters and more.

This text provides a new insight into present patterns and future trends in the consumption of fats and carbohydrates and reviews their medical significance. Considerable attention is given to the technology of modification of fats in foods and the functions of carbohydrates in food design and processing. The reader will come away with a better understanding of the role these nutrients play in the technology of food processing.

The editors wish to express their appreciation to the members of the AMA-Food Industry Liaison Committee who contributed so much to the planning and organization of the symposia on Nutrients in Processed Foods. Mrs. Margarita Nagy, Nutritionist in the Department of Foods and Nutrition, and James L. Breeling, formerly a member of the Department, but now Director of the AMA's Department of Scientific Assembly, carried the major staff responsibilities for the symposia. The editorial and technical aid of Mr. Daniel Schub and Ms. Mary Ellis is most gratefully acknowledged.

We also express appreciation for the guidance and understanding of Publishing Sciences Group, Inc.

Philip L. White, ScD
Dean C. Fletcher, PhD

This symposium on vitamins and minerals in processed foods was co-sponsored by the Council on Foods and Nutrition of the American Medical Association and the AMA-Food Industry Liaison Committee. The AMA Council has been interested for many years in the addition of nutrients to food and the provision of the best possible diet for this country. We have issued policy statements from time to time concerning fortification and enrichment and have held a number of symposia in which many of the problems of enrichment and fortification have been discussed.

The purpose of this symposium was to examine numerous aspects of vitamins and minerals in processed foods. Biological availability and requirements of some of these micronutrients were considered, as was evidence of deficiencies of these nutrients in some segments of population. Some of the factors that influence the content of vitamins and minerals in foods were discussed. The important question of how to design protective guidelines and effective food fortification programs to meet needs that are known to exist received extensive consideration. On the third day of the symposium, task force deliberations were conducted with maximum participation. The task forces developed recommendations which were presented at a final summary session. Dr. Philip Aines, chairman of the AMA-Food Industry Liaison Committee, was general chairman of the task forces.

This symposium represented a continuation of a joint program that was started three years ago by the Council on Foods and Nutrition and the AMA-Food Industry Liaison Committee. At that time, we began to bring together clinical nutritionists, nutrition scientists, food industry representatives, and scientists from the regulatory agencies to discuss problems of mutual interest. We first talked about food standards in 1969. At times we agreed, and at other times we disagreed, but we learned to know one another better and found that such meetings helped everyone in the long run. It is the co-sponsors' hope that the bringing together of the various groups in New Orleans will again prove helpful to all of us in developing recommendations for rational guidelines and nutrition policies.

Grace A. Goldsmith, MD

requirements for micronutrients in general population

PART ONE

INTRODUCTION TO PART I
by William J. Darby, MD, PhD

Some might wonder at the multiplicity of discussions such as those
we've engaged in. It might be useful to note very briefly some of
the evolution that has occurred in thinking relative to nutritional
quality of food. We became interested in the subject of enrichment
about the time of the Second World War and the years just before.
Enrichment was then a very relative concept which was based on
the presence of a demonstrated health need within the population.
And we did have a demonstrated health need in terms of
recognizable disease.
In the United States at that time we had pellagra, we had some beriberi, and we had a good deal of nutritional anemia. It was decided
that we could add nutrients to appropriate foods which would
distribute that nutrient to the segment of the population in need.
We also decided that nutrients should be available, although I'm
not sure we always got final evidence on that source, and that the
nutrients should be added only to certain foods in order to avoid
unnecessary and even promiscuous distribution.
Subsequently, there were efforts to review these guidelines and I
think, perhaps, the real change began to occur with the first
revised joint statement of the Council on Foods and Nutrition of
the American Medical Association (AMA) and the Food and
Nutrition Board. This statement on addition of nutrients to foods
dealt with the use of industrially produced nutrients in improving
the quality of the diet. The statement addressed itself to the use of
industrially produced nutrients in formulated or composite food.
Now the Food and Drug Administration has asked the Food and
Nutrition Board to provide guidelines for nu rients quality of
certain kinds of foods—convenience foods, if you will—such as
prepared meals. Our thinking is again in evolution.

In discussing the results of the 10-State Nutrition Survey as related to this symposium, we should understand that these data are preliminary. Most of the findings obtained in the first five states (Texas, Louisiana, Kentucky, Michigan, and New York) were presented at the American Public Health Association annual meeting in Houston, 27 October 1970, by Dr. A. E. Schaefer. The data pertaining to the last five states (South Carolina, West Virginia, California, Washington, and Massachusetts) and New York City were summarized specifically for this workshop.

The survey was directed toward lower income groups, based on average per capita income by 1960 census tract areas (enumeration districts) classified as urban, semiurban, and rural in each state. The sample within each state was selected from those enumeration districts with mean per capita income falling in the lower quartile. From these geographical areas, approximately 2,000 homes were chosen; they were picked from randomly selected clusters of homes and all people in every third house were included. The 10-State Survey involved a few more than 70,000 individuals; 33,500 of these from 9,000 households in the first five states. The survey consisted of a detailed clinical examination, biochemical analyses of blood and urine specimens, dietary intake records for infants (0 to 36 months), children (10 to 16 years), persons over 60 years, and all pregnant and lactating women. Patterns of food consumption were determined for households.

The survey population of the first five states can be described as 41 percent nonwhite and 59 percent white, on the average. Some 13 percent are Spanish-speaking, essentially all from the Texas sample. In terms of income, 40 percent of the families fell below the Orshansky poverty index of $3,334 for a family of four; however, this ranged from 17.8 percent in New York to 60.7 percent in Texas (Fig 1). Of the total sample, 70 percent had incomes less than $5,000 per household. Family size averaged from 4.4 to 5.7 persons.

In addition, the families that form the bulk of the sample had relatively limited education, lived in crowded housing conditions generally, and received very limited medical and dental care. At

chapter 1
evidence for deficiency in the United States

by
Gerald F. Combs, PhD

the time of the survey, only a few families participated in either the food commodity or stamp program. The percent of the persons interviewed in the homes that were examined ranged from 37 to 65 percent in the first five states, averaging 50 percent. It should be emphasized that the observations cannot be extrapolated to the entire population of any state; rather, the data should be considered applicable to the particular low-income segment of the entire population sampled.

Poor growth of children was observed in several of the states. With regard to height, the percent of children who fell one standard deviation below the mean of the Iowa growth standard in the first five states ranged from 20.3 percent to 46.5 percent (Fig 2). Valuation of bone maturation by hand-wrist radiographs also sug-

3

gests that skeletal maturation was delayed in children during the first eight years of life.

Detection of specific clinical lesions has been infrequent; however, evidence of prior rickets and enlarged thyroid has been observed. In one state, 6.6 percent of pre-school children showed clinical signs associated with prior or present vitamin D deficiency, ie, bowed legs, frontal bossing of the skull, beading of the ribs, and epiphysial enlargement. Problems such as cheilosis, angular scars of the mouth, filiform atrophy, and fungiform hypertrophy of the tongue were observed in from 0 percent to 21 percent of pre-school children and from 0.1 to 35 percent in persons over six years in the five states. Of 15,257 persons examined in the first five states, an average of 3.2 percent (1.2 to 4.9 percent) had enlarged thyroid glands (Table 1). This was not considered due to a simple iodine deficiency, except in two states where 2.6 to 5.7 percent of the urine samples showed low levels of iodine excretion. (See Table 1.)

Approximately 20 percent of the entire survey population had hemoglobin (and hemotocrit) levels which fell below those considered to be acceptable by the advisory committee. This was consistently true for both sexes and all age groups although there was considerable variation from state to state (Table 2). There was a higher incidence of unacceptable hemoglobin levels in individuals from households below poverty than from those above the poverty level. (See Table 2.) Despite this, there was a very large proportion of individuals in the above poverty group which had unacceptable hemoglobin levels. The results of the first five states surveyed showed a considerably higher prevalence of unacceptable hemoglobin levels in the Negro and Spanish-speaking groups than in the white, except for Louisiana where a very high prevalence of unacceptable hemoglobin levels was found in the whites as well (Fig 3).

Serum iron levels were considered low when they fell below 60 micrograms per 100 ml of serum in the adult male or below 40 micrograms in the adult female (Table 3). Serum iron and iron-binding capacity (transferrin saturation) were determined on blood samples, especially those having low hemoglobin levels (Tables 3 and 4). Approximately 40 percent of the population with unacceptable hema-

tocrit levels in the first five states had unacceptable serum iron or serum iron-capacity, supporting the view that an inadequate dietary intake of iron is the major cause of the low hemoglobin and hematocrit values observed in the survey population. Transferrin saturation values falling below 20 percent for adult females and 15 percent for adult males were considered to be low (Table 4).

Dietary information indicates that iron intakes were marginal to inadequate in virtually all age groups and in all states studied in the survey (Fig 4 and Table 5). The dietary intake data are based on 24-hour recalls for the various categories of individuals or households. Although one expects some variation from day to day, to have such a high proportion of individual intakes failing to meet 70 percent of the dietary standard (NRC allowance for iron) supports the biochemical data that indicate a widespread iron insufficiency in the survey population. (See Table 5.) Red cell folate values together with the percent of males and females falling below 140 or 160 nanograms per ml are given in Table 6. Values falling below these levels are considered to be unacceptable. It should be remembered that most of these values were obtained from blood samples showing low hemoglobin values, which may account for the relatively high percentages of unacceptable folates. It seems clear, however, that the possibility that folate deficiency is involved in the widespread anemia problem is most real. Perhaps up to 10 percent of the low hemoglobin levels are due to folate deficiency. (See Table 6.)

Another widespread finding in the 10-State Nutrition Survey was the high prevalence of low plasma vitamin A levels (Table 7). Since these data were summarized, it has been determined that the data from New York State were expressed in terms of vitamin A acetate rather than vitamin A alcohol. Accordingly, the mean value is approximately 13 percent higher than the true mean and, accordingly, the percent of individuals falling below 10 or 20 micrograms of vitamin A alcohol per 100 ml is inordinately low. (See Table 7.)

Even after these corrections are made, however, the incidence of low serum vitamin A's in New York will be relatively low. Further examination of the low serum vitamin A levels shows that most of these are found in individuals under 17 years of age. Table 7 compares the prevalence in the 10- to 12-year-old groups versus the total sample population and does reveal a higher proportion in that age group. Except for certain states (Kentucky, Texas, and Michigan), there is little difference in the incidence of low serum vitamin A's in white and nonwhite population groups.

Dietary intake data indicated that✓ from 20 to 40 percent of the sample population is consuming inadequate amounts of vitamin A. Frequency of consumption of vitamin A-rich foods indicated that approximately 45 percent of the families in the first five states seldom or never consumed foods considered to be good sources of vitamin A. Clinical signs associated with inadequate vitamin A intake in the first five states surveyed revealed only 23 persons with Bitot spots, but 404 persons exhibited follicular hyperkeratosis out of the 15,257 persons examined.

The low serum vitamin A's observed in these studies are very similar to other unpublished findings of Hepner at the University of Maryland in young children and by Sandstead, Vanderbilt University, among pre-school children enrolled in Headstart. While clinical evidence does not suggest that this has yet led to severe vitamin A deficiency disease, the widespread occurrence of low serum vitamin A levels warrants preventive efforts to insure that this prevalence of low vitamin A levels in children does not persist.

Considerable variation occurred with respect to the levels of serum vitamin C considered to be below the acceptable standard (Table 8). The findings suggest little relationship between age or ethnic groups in relation to vitamin C levels. Poverty had relatively little effect on the occurrence of low vitamin C levels. Of the population groups surveyed, the Negroes and poor whites in Kentucky, West Virginia, Texas, and Louisiana have the highest prevalence of unacceptable serum vitamin C levels.

The dietary information supports this regional difference in the prevalence of low serum vitamin C levels, in that the dietary intakes were considerably lower in these states. Unacceptable riboflavin excretions (Table 9) ranged from 4 to 30 percent in the various states, with the highest prevalence of low excretions in relation to creatinine excretion falling in the under-16-year group (Fig 5) and in minority group individuals, particularly the Negro and Spanish-Americans. Individuals living below the poverty level generally had a slightly high prevalence of low riboflavin excretion.

Thiamine excretions, also shown in Table 9, were not considered to represent an appreciable problem although 10 to 12 percent of the individuals in West Virginia, New York City, Texas, and Louisiana showed excretion levels below that considered satisfactory. Problems of methodology with thiamine, however, would suggest that this is of minor importance at this time.

This presence of multiple unacceptable biochemical values can be considered one prime indication of nutritional risk. In a well-fed population, one would not expect to find many persons with multiple unacceptable biochemical values. From 1.0 to 16.4 percent of the population surveyed had two or more unacceptable values out of four (Fig 6). In terms of age groups, children under 16 had the highest rates of multiple unacceptable values.

Here, a clear relationship exists between income and nutritional status, but it should be remembered that there still is a significant prevalence of unacceptable values in the higher income groups. In the survey population, the minority groups, especially the Negro and Spanish-speaking, tended to have higher prevalences of unacceptable values. At the same time, they tended to be concentrated in the lowest economic groups, and at this point it is very difficult to separate the effects of ethnic group and income with respect to these data.

The problems of nutrition must be considered in the context of *socioeconomic and cultural factors, education, foods consumed, and general health.* If we are to alleviate and eradicate malnutrition, it is necessary to improve *more than just the food patterns of the vulnerable population groups.* Programs must be undertaken to:

✓ 1. *Improve the nutritional quality of key staple foods by up-dating food enrichment standards.*

2. *Monitor in-depth the impact of our changing food supply and environment.*

3. *Educate the public regarding proper nutrition and infant and child feeding.*

4. *Bring the vulnerable population groups into the welfare, food, education, and health care delivery systems.*

Such programs must be aimed initially at the poverty groups, but will in time spill over into other segments of the population where problems of nutrition also exist.

Table 1.
Thyroid Enlargement and Urinary Iodine Excretion (Both Sexes).

Survey	% Thyroid Enlarge.	Urinary Iodine, μg/g Creatinine Median	<50
Tex.	4.3	408	0.3
Ky.	4.9	310	5.7
Mich.	1.2	312	2.6
N.Y.	2.7	213	1.9
Mass.	N.A.	267	3.1

Table 2.
Comparison of Number, Percent Deficient, and Percent Low Between Sexes for Hemoglobin; 10 States and New York City Nutrition Surveys, 1968-70 (Preliminary).

	Hemoglobin, gms % (All Males)				(All Females)			
Survey	(No.)	Mean	% Def.	% Low	(No.)	Mean	% Def.	% Low
Tex.	(1378)	13.3	4.6	21.2	(1880)	12.7	7.0	17.0
Ky.	(580)	13.3	5.5	23.3	(702)	12.8	2.3	15.5
Mich.	(803)	13.5	3.0	21.7	(1114)	12.9	1.2	16.6
N.Y.	(1318)	14.0	2.3	13.0	(1493)	13.2	0.9	8.2
LA.	(1874)	12.4	NA	42.0	(2671)	12.0	NA	35.7
Wash.	(929)	14.2	0.6	10.5	(1136)	13.3	0.9	7.2
Calif.	(2023)	14.2	2.0	11.9	(2767)	13.4	1.2	9.2
S.C.	(1559)	12.6	9.9	42.7	(2193)	12.0	6.7	32.8
Mass.	(1591)	14.2	1.8	10.6	(1949)	13.4	0.9	7.7
W. Va.	(542)	13.8	2.0	17.0	(756)	13.2	1.1	9.6
N.Y.C.	(789)	13.6	2.8	17.1	(1096)	12.7	2.3	16.8

Hemoglobin Deficient and Low Standards (gm/100 ml)			
	Deficient	Low	Acceptable
6-23 months	< 9.0	9.0- 9.9	≥10.0
2-5 years	<10.0	10.0-10.9	≥11.0
6-12 years	<10.0	10.0-11.4	≥11.5
13-16 male	<12.0	12.0-12.9	≥13.0
13-16 female	<10.0	10.0-11.4	≥11.5
>16 male	<12.0	12.0-13.9	≥14.0
>16 female	<10.0	10.0-11.9	≥12.0
Pregnant, 3rd Trimester	< 9.5	9.5-10.9	≥11.0

Table 3.
Serum Iron, μg/100 ml (10-State Nutrition Survey).

		(Males)				(Females)	
Survey	(No.)	Mean	% Low	(No.)	Mean	% Low
Ky.	(192)	83.2	21.4	(234)	73.2	20.5
Mich.	(118)	79.5	23.7	(104)	68.5	18.3
N.Y.	(94)	76.5	30.9	(44)	67.7	27.3
Wash.	(838)	93.6	11.6	(1005)	87.9	6.7
Calif.	(22)	84.8	4.5	(42)	79.8	16.7
S.C.	(281)	68.1	36.3	(256)	63.3	28.6
Mass.	NA	NA	NA	(95)	99.6	4.2
W. Va.	(37)	83.5	13.5	(24)	75.3	16.7
N.Y.C.	(380)	92.0	13.5	(457)	83.3	6.6
Tex.	(222)	84.0	12.8	(264)	74.8	17.2

Table 4.
Transferrin Saturation, Percent (Males) [10-State Nutrition Survey].

Survey	(No.)	Mean	% Low
Ky.	(180)	26.2	27.3
Mich.	(34)	20.9	29.4
N.Y.	(93)	29.5	38.7
Wash.	(742)	28.2	20.9
S.C.	(95)	19.2	47.4
N.Y.C.	(305)	26.3	20.6
Tex.	(200)	24.5	37.0

Table 5.
Calculated Mean Dietary Iron Intakes (mg/day) [10-State Nutrition Survey].

Survey	House-holds	0-3 Yrs	10-16 Yrs	>60 Yrs	Preg. & Lact.
Tex.	13	6	14	11	16
La.	12	8	12	10	10
Ky.	13	8	12	9	10
N.Y.	14	10	14	12	13
Mich.	14	8	13	12	12

Table 6.
Red Cell Folate, ng/ml (10-State Nutrition Survey).

		(Males)					(Females)		
Survey	(No.)	Mean	% < 140	% < 160	(No.)	Mean	% < 140	% < 160
Ky.	(427)	266.0	13.5	21.5	(543)	260.	13.1	18.4
Mich.	(220)	155.0	53.1	61.6	(308)	167.	49.1	57.3
N.Y.	(463)	295.0	10.9	14.1	(503)	281.	9.0	13.4
Wash.	(121)	250.0	27.1	30.0	(127)	257.	15.3	21.1
Calif.	(163)	245.0	20.3	29.3	(259)	235.	15.6	27.1
S.C.	(506)	202.0	29.8	38.9	(729)	207.	25.3	35.2
Mass.	(892)	226.0	17.8	26.6	(1087)	227.	17.8	25.6
W. Va.	(253)	299.0	7.0	9.7	(295)	296.	7.3	10.3
N.Y.C.	(389)	217.0	22.4	30.9	(513)	230.	19.7	27.7
Tex.	(131)	206.3	25.2	35.1	(206)	208.	27.7	33.0

Table 7.
Serum Vitamin A, μg/100 ml (10-State Nutrition Survey).

(All Samples)					(10- to 12-yr.-olds)				
Survey	(No.)	Mean	%<10	%<20	Survey	(No.)	Mean	%<10	%<20
Tex.	(370)	25.8	2.2	25.4	Tex.	(2997)	31.6	1.6	16.8
Ky.	(111)	32.0	6.3	11.7	Ky.	(1046)	40.1	3.6	9.3
Mich.	(198)	33.1	.0	4.5	Mich.	(894)	40.4	0.1	4.2
N.Y.	(80)	47.4	.0	2.5	N.Y.	(1055)	65.0	0.3	1.3
Wash.	(152)	26.8	6.6	26.3	LA.	(3898)	47.2	0.1	2.3
Calif.	(371)	44.0	.0	3.0	Wash.	(1348)	31.1	5.4	23.5
S.C.	(418)	29.3	.0	8.4	Calif.	(3717)	57.8	0.1	1.2
Mass.	(278)	36.9	0.6	15.5	S.C.	(2158)	34.1	0.6	9.6
W. Va.	(110)	30.8	.0	15.5	Mass.	(2503)	44.4	1.0	10.2
N.Y.C.	(169)	36.3	0.6	5.3	W.Va.	(643)	37.7	.0	7.6
					N.Y.C.	(1096)	44.9	0.5	3.7

Plasma Vitamin A (μg/100 ml) Deficient and Low Standards

	Deficient <10	Low 10-19	Acceptable $\geqslant$20
All ages			

Table 8.
Serum Vitamin C, mg/100 ml (Total Samples) [10-State Nutrition Survey].

(Total Samples)			
Survey	%<0.2	Survey	%<0.2
Wash.	2.0	Tex.	12.1
Calif.	2.9	La.	14.0
S.C.	2.5	Ky.	8.0
Mass.	5.7	Mich	0.4
West Va.	7.6	N.Y.	2.4
N.Y.C.	0.8		

Serum Vitamin C (mg/100 ml)
Deficient and Low Standards

	Deficient <0.1	Low 0.1-0.19	Acceptable $\geqslant$0.2
All ages			

Table 9.
Urinary Excretion, μg/g Creatinine (10-State Nutrition Survey).
Total Sample

Survey	Riboflavin % Low & Def.	Thiamine
Wash.	11.	6.
Calif.	7.	5.
S.C.	30.	9.
Mass.	7.	4.
W. VA.	10.	12.
N.Y.C.	9.	11.
Tex.	21.	10.
La.	15.	11.
Ky.	10.	6.
Mich.	13.	6.
N.Y.	4.	4.

Urinary Riboflavin Deficient and Low Standards (μg/gm creatinine)

	Deficient	Low	Acceptable
1-3 years	< 150	150-499	≥ 500
4-6 years	< 100	100-299	≥ 300
7-9 years	< 85	85-269	≥ 270
10-15 years	< 70	70-199	≥ 200
Adult	< 27	27-79	≥ 80
Pregnant 3rd Trimester	< 30	30-89	≥ 90

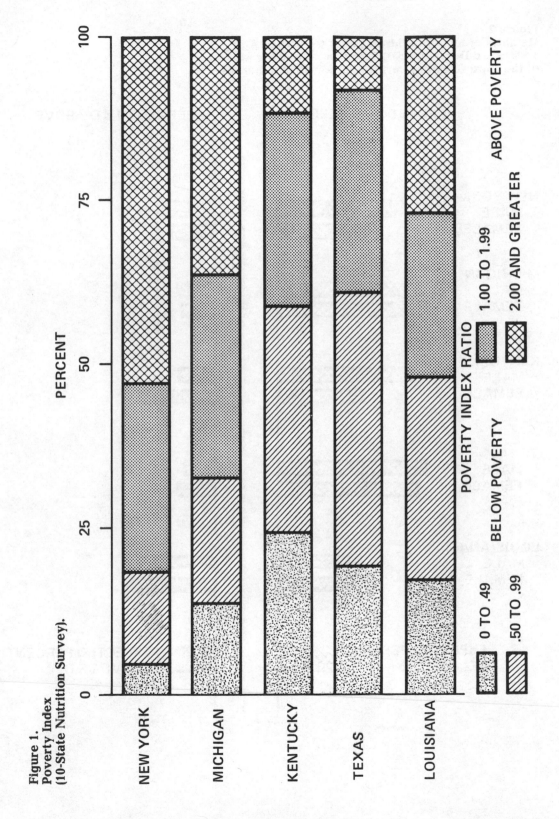

Figure 1.
Poverty Index
(10-State Nutrition Survey).

PERCENT

POVERTY INDEX RATIO

BELOW POVERTY   ABOVE POVERTY

0 TO .49   1.00 TO 1.99

.50 TO .99   2.00 AND GREATER

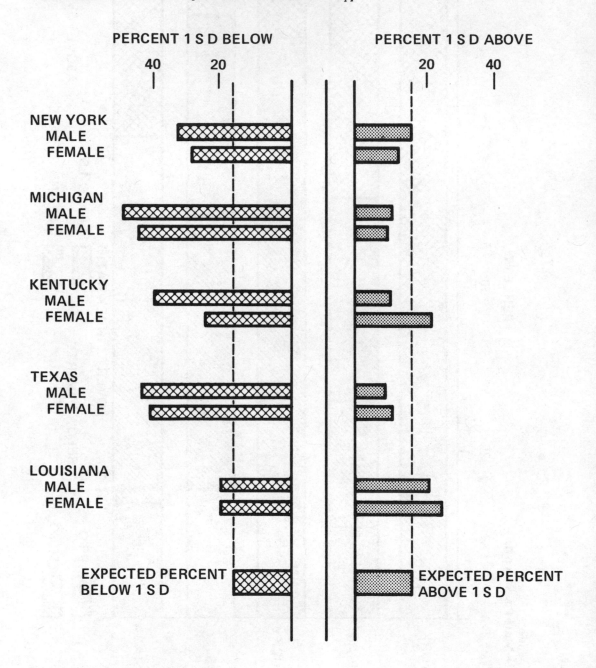

Figure 2.
Height: Percent of Children Under 6 Years of Age
1 Standard Deviation (SD) Below or Above the Mean
of the Iowa Growth Chart [10-State Nutrition Survey].

Figure 3.
Hemoglobin: Percent Unacceptable by Ethnic Group
(by State) [10-State Nutrition Survey].

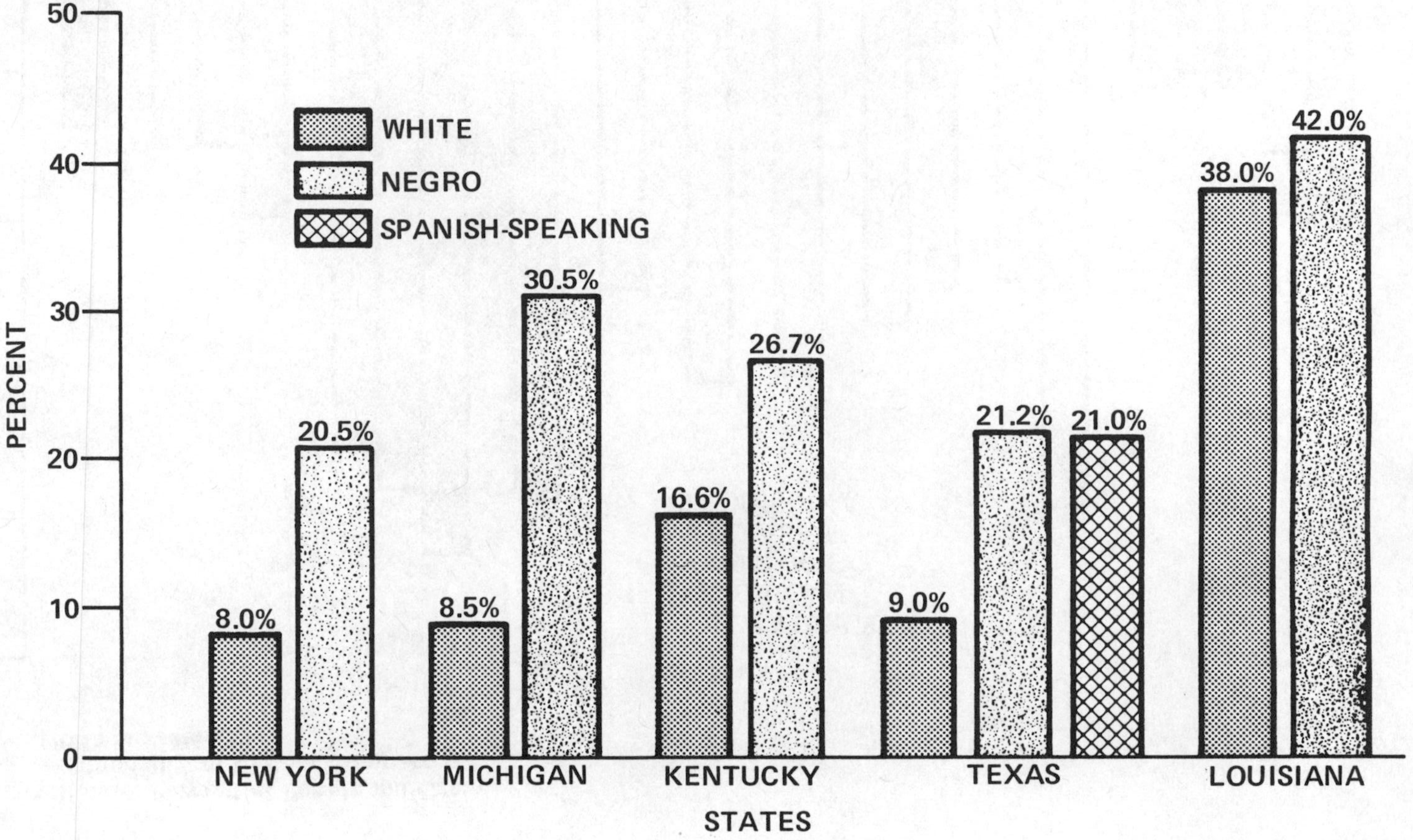

13

Figure 4.
Iron–Dietary Intake: Percent of Population Groups
Consuming Less than 70% of Dietary Standard
(10-State Nutrition Survey)

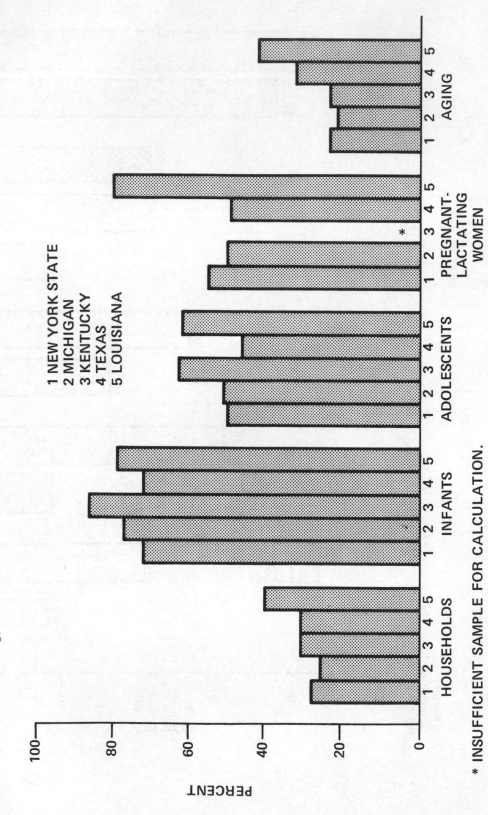

1 NEW YORK STATE
2 MICHIGAN
3 KENTUCKY
4 TEXAS
5 LOUISIANA

PERCENT

* INSUFFICIENT SAMPLE FOR CALCULATION.

14

Figure 5.
Urinary Riboflavin Unacceptable by Age
(by State) [10-State Nutrition Survey].

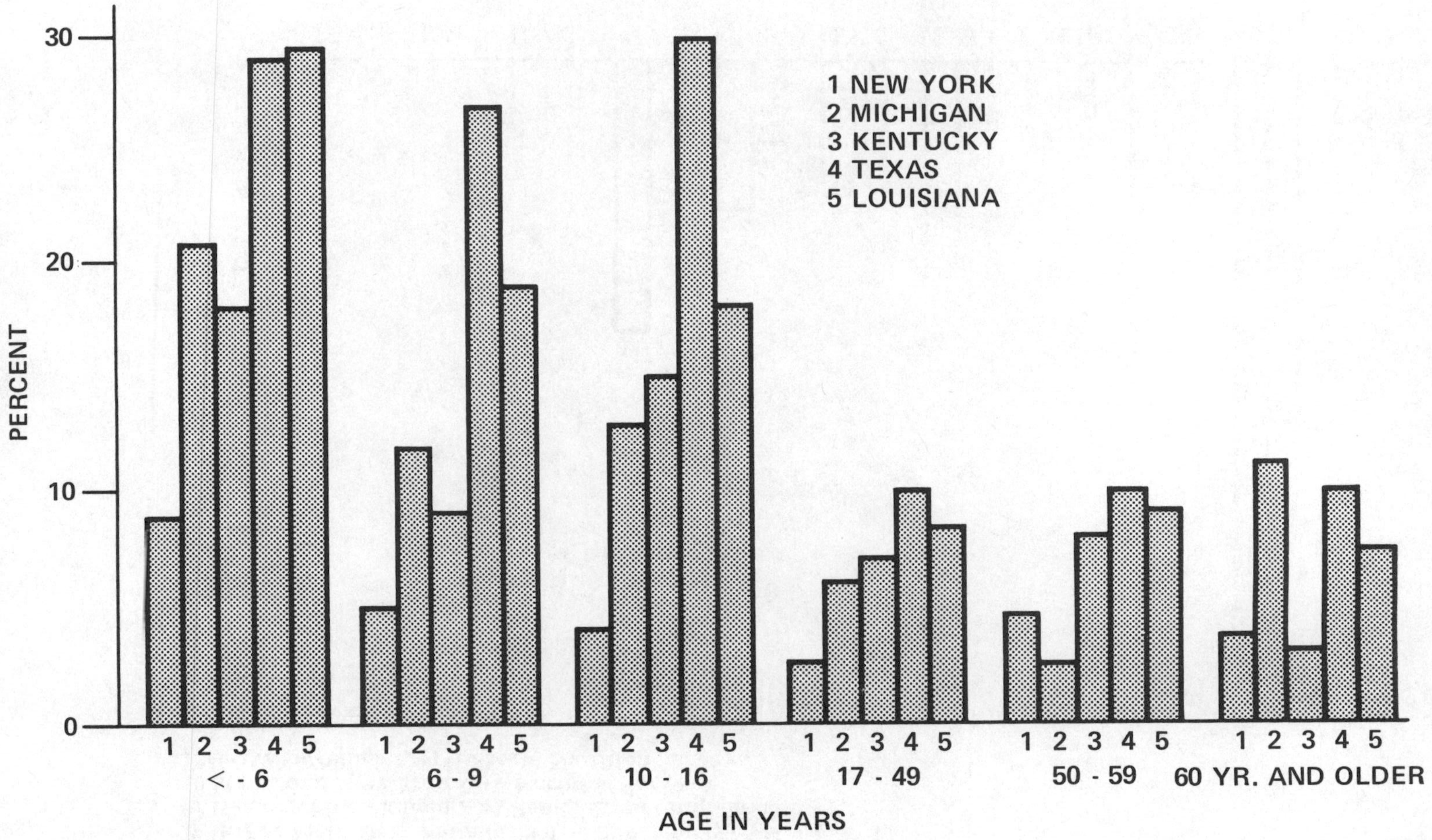

30
20
10
0
PERCENT
1 NEW YORK
2 MICHIGAN
3 KENTUCKY
4 TEXAS
5 LOUISIANA
1 2 3 4 5
< - 6
6 - 9
10 - 16
17 - 49
50 - 59
60 YR. AND OLDER
AGE IN YEARS

Figure 6.
Percent of Persons in the Below and Above Poverty Groups
With 2 or More Deficient and/or Low Biochemical Values
in Hemoglobin, Vitamin A, Vitamin C, and Riboflavin by
10 States and New York City Nutrition Surveys
1968-1970 (Preliminary) [10-State Nutrition Survey].

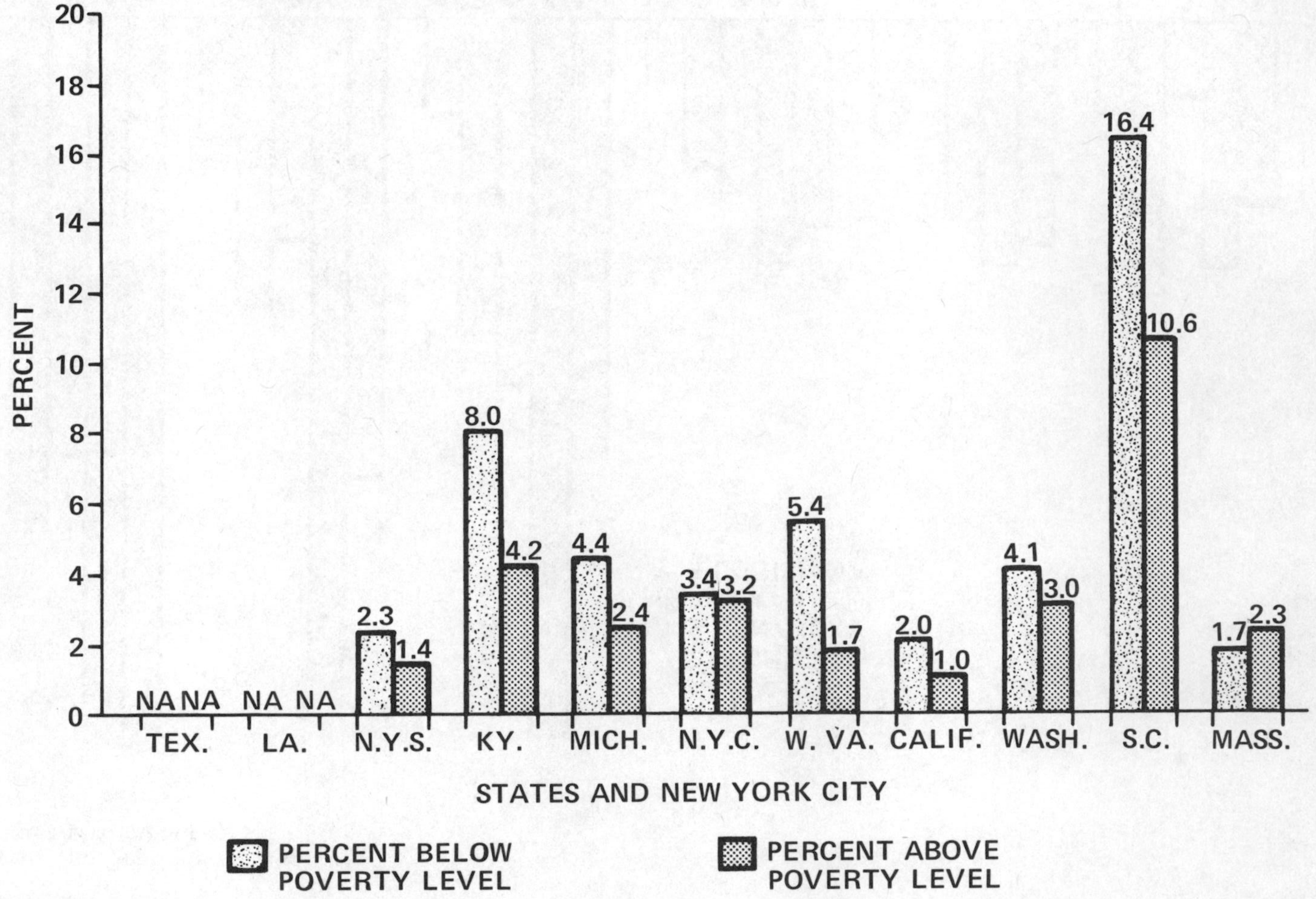

My subject calls for heavy emphasis on things that we don't know. In order not to be misunderstood, I would like to emphasize at the beginning that I'm a proponent of enrichment and fortification. I do believe we need it badly.

I have drawn up some data from the American Institute of Baking where the content of trace elements in milling fractions was compared with the content originally present in whole wheat and I have plotted the percent of the initial concentration remaining in flour. We all know that iron goes down to a little over 20 percent of its initial values. This is also true for a whole list of other essential elements—for example, zinc, manganese, copper, molybdenum, and cobalt.

We are not justified in saying we don't have to worry about this reduction because these elements are poorly available due to phytate or phosphate contents in the whole wheat fractions. In the absence of availability data for *all* essential trace elements, we must be concerned with this loss and its consequences for our daily intake.

Another problem is perhaps even more urgent. We can predict a continuing and perhaps increasing trend toward the consumption of formulated products in which we substitute an isolated vegetable protein for meat. These may have a very good protein composition but—unless we add the trace elements needed—we will have an inferior product. If the consumption of these products increases to a significant proportion of our total intake, we can almost certainly predict deficiencies unless we enrich to levels similar to those in the food that the new product is intended to replace.

In his introduction, Dr. Darby discussed severe deficiencies, such as pellegra, rickets, and scurvy, which, thanks to nutrition research, have disappeared in this country. He could have drawn a theoretical distribution curve of the population at that time that showed two clearly different peaks, one representing a severely deficient population and one a normal population. This pattern, we all agree, has disappeared.

We are now dealing with a situation where this "deficiency peak" has shifted more and more toward the "normal" one and, as far as iron is concerned, has even shifted into it so that it forms a hump on one side. In other words, we are not

chapter 2
micronutrients needing better definition of requirements

by
Walter Mertz, MD

concerned with severe deficiencies any more. We are concerned with sub-optimal, marginal intakes, and sub-optimal functions.

This presents a problem because it raises the question—how does it hurt you if you do not have a hemoglobin level of 16 or 15? Do you really undergo any risk? This doesn't only apply for iron, but it will soon be an important problem with regard to zinc and other elements. We have to be aware that we have to balance one risk of perhaps giving too much to certain population groups against the risk of marginal deficiency.

What is marginal deficiency? We have seen that a substantial number of people have an undesirably low hemoglobin concentration. We know that a marginal intake of zinc can lead to impaired wound healing. We know that a

19

marginal intake of chromium can lead to impaired glucose metabolism. But now, we have to ask ourselves, does that really hurt you? I believe it does.

While there is no question about impaired wound healing, there is controversy concerning marginal iron deficiency and little information about the long-term consequences of impaired glucose tolerance. But I firmly believe that we should try for optimal intake and for optimal function. There are many people who are concerned about excessive intakes of micronutrients.

There is the emotional controversy with regard to fluoridation of water, and even the iodization program has had many adversaries. There are scientists who don't believe in enriching with iron because it creates certain risks to a small number of people. What we have to do in order to agree on a requirement is to define as closely as we can the long-term consequences of a sub-optimal intake.

One of the most interesting and promising developments of the past few years is the demonstration by two different groups that a marginal iron deficiency will result in a diminished ability of children to pay attention to what they are being taught. These data are just beginning to be significant and this is a very important beginning.

The apparent incidence of marginal zinc deficiency is much more difficult to determine by analytical or functional means. However, it appears that there are many people who benefit from an additional zinc intake once they are subjected to wound or burn stress in the hospital. How can we catch these people before they get to the hospital? I am often asked if you have chromium deficiency in people and they have an impaired glucose tolerance, how does it hurt them? I don't know the answer.

I know that I would much rather have a normal glucose tolerance than an impaired one. But here again, we lack knowledge of the long-term consequences. When trying to determine this marginal state, we must realize that chemical determination of the micronutrients per se does not necessarily give us meaningful information. We are determining a total of which only a small part might be meaningful. This is certainly true for chromium and probably true for several others.

A plasma chromium level in the fasting state does not give any information as to the nutritional status. It reflects the recent dietary intake. If you challenge a person who is chromium deficient with a dose of insulin or glucose which causes an increase of meaningful chromium fraction in the plasma, then you have a way of showing whether this person has enough reserves of this special chromium or not. We have seen an elderly person who had absolutely no increase of plasma chromium following a glucose load indicating that the chromium pool was empty. After several weeks of trace supplementation with chromium, the tests were repeated, and then this person was able to produce a significant increase of chromium in the plasma.

Similar criteria might be developed for a number of other trace elements and we do know a number of criteria that can be used. Let me just remind you of transketolase activity and thiamine, of alkaline phosphatase and zinc, transferrin saturation and iron, and the above example of glucose challenge with regard to chromium.

Before discussing the category of trace elements recently shown to have an essential role, I must state that we should never be satisfied that we know all the essential micronutrients. This is certainly true for trace elements and quite probably true for the vitamins. I believe that it is not only possible but likely that more will be discovered in the future.

I will discuss selenium first. There is no doubt that selenium is essential in the various species in which it has been tested. There is suggestive evidence that it has a function in malnourished children and that it can speed up recovery when the selenium levels are initially low. There is evidence from Dr. Allaway's studies that there are considerable regional differences in the blood selenium levels of man in this country. What these mean we don't know. There has been work about the possibly beneficial role of selenium supplements in cancer. There is also work that says exactly the opposite.

We have come to a point where feed manufacturers are getting concerned about selenium deficiency, and past history tells us that this may mean that in a few more years we have to become concerned with deficiency in humans. This is the way that the zinc story has gone. We do not know what the selenium requirement is, but we need to know it exactly because the margin of safety of selenium is relatively small.

The next element is zinc. I am giving just one example from Pories' work where the rate of wound healing was determined in controls and in people who got extra zinc supplement. Around day 15 of supplementation, a very dramatic increase in the healing rate began in the zinc-supplemented subjects. As I said before, zinc deficiency is not uncommon in animals and we have now learned to supplement animal feeds routinely with this trace element. The requirement in humans which is, according to balance studies, anywhere from 8 to 10 milligrams per day is met by good diet, but there is not a wide enough margin of safety so that I feel we will have to study this question more carefully.

The next element is chromium. Chromium is a co-factor for the hormone insulin. It is relatively difficult to induce a chromium deficiency in animals but it can be done. To our surprise, it has been shown that in the population in this country, the average tissue chromium concentration declines with age. This is unusual for a trace element because most remain steady or keep going up. This decline may be correlated in some way to the decline of efficiency of glucose utilization with age. As you grow older your glucose tolerance decreases.

In the few studies that have been done, approximately half of the subjects—middle-aged and old—supplemented with an additional 50 micrograms of chromium per day responded with improved glucose utilization. The resulting curves were very similar to those found in young adults. So, it is possible that we do have a nutritional problem with chromium in this country.

To estimate the chromium requirement of man is difficult for two reasons. Firstly, the analytical problems for this element have not yet been completely solved; therefore estimates of daily intake and of chromium balance are not yet reliable. Secondly, the biological availability of different chromium compounds varies widely, and the best available compounds, such as occur in foods, have not been identified as yet and cannot be measured directly. Whether the estimated average intake from US diets of 60 μg/day

is optimal or even sufficient is a matter of conjecture.

I will now turn to three micronutrients that have emerged during the past year as having a biological function. The first one is tin. Dr. Schwarz discovered that rats raised in a strictly controlled environment and fed an amino acid diet respond to supplementation with tin with a very significant growth increase. The levels that are required for growth stimulation are around one part per million in the diet.

The second micronutrient is nickel. Dr. Forrest Nielsen in our laboratory has been able to produce nickel deficiency in the chicken. The symptoms consist of a slight swelling of the hock and a different pigmentation of the leg skin. This syndrome appears with diets that contain less than 80 parts per billion of nickel and it is prevented by adding nickel back to the diet. With this are correlated some metabolic changes which apparently have to do with lipid metabolism but they have not been completely worked out yet.

The third micronutrient is vanadium. Dr. Leon Hopkins in our division, using the same controlled-environment isolator system and highly purified diets, was able to produce vanadium deficiency in the chick which is manifested by an impairment of feather growth. Now, what are the implications of these findings for man? If it is so difficult to produce a deficiency, is it not unlikely that deficiency states would spontaneously develop in man?

We said exactly the same thing for chromium deficiency. Chromium deficiency is difficult to produce and yet there is mounting evidence that we may be facing a real nutritional problem in the population. Therefore, these data should alert us to the possibility that nutritional problems for any one of these elements may emerge at some later time even in man.

When discussing micronutrient requirements, we must realize that one trace element or vitamin never exists by itself in a diet and it is well known that an acceptable level of one, even if beneficial by itself, can induce deficiencies or difficulties with another. The typical example is the relationship between calcium and zinc.

Another relationship which might be very important, in the light of recent work in Ithaca on the reversibility of osteoporosis, is that of calcium and phosphate. The interaction of selenium with toxic metals like mercury and cadmium is one of the most challenging pieces of research.

Finally, let us examine the problem of availability. We realize that determination of total concentration of a micronutrient in diets may be meaningless. The chemical determination does not describe how much of this total concentration is really available to the organism. We know reasonably well some of the interactions, such as the interaction between phosphates and iron, phosphates and zinc, and phytates and several trace elements. We know that certain chelators like EDTA may be detrimental to the absorption of some trace elements, and that others, such as ascorbic acid, can significantly increase the availability.

For example, we have measured chromium in lettuce and beer, both of which have a similar chromium content when expressed on a dry weight basis, and we could say both are good sources. Beer is just one example of a substance with a good content of available chromium.

When we extracted the chromium from the material and measured it in our biological assay, on the basis of its potentiating action on insulin, the chromium from lettuce was completely ineffective, whereas the activity that we extracted from beer had a strong potentiating action on insulin. This shows that we must be concerned with the availability of trace elements in biological materials and the same is true for some vitamins.

Dr. Sweeney in our laboratory is studying the isomerization changes that the carotenes in food undergo when they are cooked or otherwise processed. We know that the all-trans carotenes remain available, whereas other categories—the Neobeta B, Neobeta U—are less available. During 25 minutes of cooking, the well-available fractions in broccoli declined somewhat and the Neobeta U—which is less available—increased.

It is important to realize: this pattern is different for individual foods. In sweet potatoes, for example, there is a great increase in Neobeta B—whereas this fraction does not increase at all in broccoli. This shows just a few of the problems that exist with regard to availability.

In summary, my main plea would be that we try to identify very clearly the consequences of marginal deficiencies of essential nutrients. Only if we can clearly define these consequences will we be able to make a strong case for restoring the levels of essential nutrients in processed foods and compensate for the losses due to processing.

As a pediatric nutritionist, examining the question of excessive intake or imbalances of vitamins and minerals in children, I will focus primarily upon the fat soluble vitamins A, D, and K; phosphorus; and the trace elements iodine, zinc, and cadmium. With the exception of the latter two trace elements, excessive intake or imbalance of these nutrients produces clinical signs and symptoms which the physician recognizes.

VITAMINS

vitamin A The Committee on Drugs and the Committee on Nutrition of the American Academy of Pediatrics have prepared a statement on the Use and Abuse of Vitamin A in which they recommend: (1) that the sale of high potency vitamin A preparations be restricted to prescription; and (2) that the physician develop awareness of the vitamin A content of preparations prescribed.[1]

Vitamin A toxicity in infants as manifest by a bulging fontanel, irritability, and an increase in cerebral spinal fluid pressure has been described on daily doses of 22,000 International Units per day. The vitamin preparation used was of a water soluble form and the duration of dose was 4 to 6 weeks. In older children, primarily adolescent girls with acne, doses of 25,000 to 50,000 IU per day will produce a state of hypervitaminosis which simulates the signs and symptoms of an intercranial neoplasm. Thus, when one exceeds the recommended daily allowance of infants and children by a factor of 10, a toxic state may ensue.

What is the significance of this statement on vitamin A to the food industry? This is best illustrated by a question raised by a pediatrician who noted in his practice that it was not unusual for young children to eat one or more jars of carrots, squash, or sweet potatoes daily. Food composition tables supplied by infant food manufacturers indicate that a single jar of carrots may supply as much as 28,000 International Units of vitamin A. A jar of sweet potatoes may supply 9,000 International Units. If vitamin A is toxic at these levels of intake, what advice should the pediatrician be giving to mothers? The answer to his question is reasonably simple in that the vitamin A data released by the infant food processor is in error. In converting carotenoid pigment to units of

chapter 3
excessive intake and imbalance of vitamins and minerals

**by
Lloyd J. Filer, Jr.
MD, PhD**

preformed vitamin A, a 1:1 conversion was used rather than the more appropriate 1:6 for beta-carotene or 1:12 for other carotenoids.[2]

Thus, food tables listing vitamin A equivalents from carotenoids need to be revised to bring the data into accord with facts. To properly inform the physician and consumer, a more accurate statement of vitamin A equivalent is required.

Do we require more vitamin A in the dietary of infants and children? In the Preschool Nutrition Survey of Owen only 2.5 percent of 596 children 12 to 72 months of age studied in the first half of the sample were found to have serum vitamin A levels less than 20 micrograms per 100 ml of plasma.[3] Ten percent of these children were found to consume a diet providing less than 100 International

Units of vitamin A per kg body weight per day. Underwood and her co-workers found that children under 10 years of age had the highest liver stores of vitamin A, 3 times greater than any other age group examined.[4] In the Underwood Study, however, sample size was small; thus, more data are needed. The situation with respect to vitamin A intake of children does not appear to be one of increasing the level or types of food fortified with vitamin A, but rather the provision, primarily to low-income families, of those foods which are currently known to provide an adequate intake of preformed vitamin A or provitamin A.

vitamin D In 1963 the Committee on Nutrition of the American Academy of Pediatrics recommended that foods other than milk and infant formula products not be enriched with vitamin D.[5] They further recommended that the daily intake of vitamin D from all sources be limited to amounts not greatly in excess of 400 International Units per day. Upon reexamination of the relationship of vitamin D to infantile hypercalcemia and the dangers of vitamin D to the population at large, they reaffirmed this position in 1965 and again in 1967.[6,7]

A survey of daily intake of vitamin D by infants and children was published by Dale and Lowenberg[8] in 1967 (Table 10). For these children, total intake ranges from 460 to 660 International Units per day. For the age range 12 to 14 years, foods other than milk then supply approximately 30 percent of total daily intake. These foods comprise primarily margarine, chocolate flavoring, and diet foods. Milk was found to be the most uniform source of vitamin D for all age groups; other vitamin D fortified foods were not needed to bring vitamin D intake to the desired level.

Infantile hypercalcemia exists in two forms, one mild and reversible in which the clinical course is marked by failure to thrive and mild azotemia. The other form is severe and irreversible wherein the infant manifests, in addition to its failure to thrive, mental retardation, over-mineralization of the base of the skull, and the metaphyses of the long bones and congenital heart disease expressed primarily as supra-valvular aortic stenosis.[9] Since the severe form was thought to arise in utero, there was an expressed concern about vitamin D intake

during pregnancy. The offspring of rabbits fed large doses of vitamin D (1.5 million units per day) throughout pregnancy were found to have vascular lesions resembling those seen in the human.

It has been reported recently, however, by Goodenday and Gordan that in 14 hypoparathyroid women taking 50,000 to 250,000 International Units of vitamin D_2 daily, 21 normal infants were delivered.[10] examination of these children at ages 6 weeks to 14 years revealed no craniofacial or cardiovascular abnormalities. These investigators concluded that massive doses of vitamin D were not toxic to the human fetal cardiovascular system.

It is clear that the recommended allowance of 400 International Units of vitamin D per day amply provides for the total vitamin D requirements of normal infants, children, and pregnant women. The suggestion that vitamin D might be a primary factor in causing infantile hypercalcemia should caution us to give careful consideration to our approach to fortification with this essential nutrient. Fortification of foods other than milk and infant formula products appears unnecessary.

vitamin K In 1946, Rapaport and Dodd reported hypoprothrombinemia in 7 infants, 4 to 12 months of age, whose primary disease was chronic diarrhea.[11] These infants had received sulfonomides during the course of their illness. Administration of vitamin K produced normal prothrombin time in these infants. These investigators considered inadequate intake of vitamin K as a possible factor in the development of hypoprothrombinemia.

In 1966, Goldman and Deposito reported observations on five infants beyond the newborn period who developed acute bleeding episodes that responded to vitamin K administration.[12] All of these infants were on a diet low in content of vitamin K. Three of the 5 infants had diarrhea and one was receiving daily doses of mineral oil. Four of the 5 infants had received a course of antibiotics.

Hypoprothrombinemia is frequently found in patients with diarrhea receiving antibiotics; however, most of these patients do not bleed. Goldman considers that the low vitamin K content of the formula fed these infants may have aggravated their deficiency state[13] (Table 11).

On the basis of these studies, Goldman considers it prudent to provide additional vitamin K to the infant receiving a low vitamin K intake who, in addition, has diarrhea and/or is receiving antibiotics. In a later report, Goldman and his co-workers described 15 additional cases of bleeding in infants beyond the newborn period.[14] Thirteen of the 15 infants had diarrhea and received antibiotics. All were on casein hydrolysate or meat-based formula. In a controlled study wherein infants with diarrhea were assigned to either a cow's milk formula or a formula whose protein source was a casein hydrolysate, 8 of 22 infants fed the latter formula became hypoprothrombinemic (Table 12). Transfer of three of these infants to a formula prepared from milk resulted in normal prothrombin times.

Based largely on this clinical evidence and recent studies which lead one to question the availability to the host of vitamin K synthesized by the microflora of the gut, two infant formula manufacturers are adding vitamin K to nonmilk-based formulas that assay below 20 micrograms per liter. The Food and Drug Administration has sanctioned this addition which should prove to be a lifesaver under certain conditions.

The Committee on Nutrition of the American Academy of Pediatrics has made the tentative recommendation that all formula products containing less than 25 micrograms per liter of vitamin K_1 should have vitamin K_1 added to attain a level of at least 100 micrograms per liter.[15] These observations on formula products, indicating that soy protein or meat-based formulas may differ from milk-based formulas as far as meeting the vitamin K requirement of the infant, raise the distinct possibility that we will need to consider the vitamin K content of foods produced from textured proteins or other more refined protein or amino acid sources.

MINERALS

phosphate The addition of phosphate to foods may be either detrimental or beneficial depending upon age. In the case of the newborn infant fed a formula of evaporated milk, water, and added carbohydrate wherein the evaporated milk has been stabilized by the addition of sodium phosphate, neonatal tetany or seizures frequently occur. Under the stan-

dard of identity for evaporated milk, it is permissible to add additional phosphate to insure a stable product during processing. This addition is such that the usual calcium to phosphate ratio of milk, 1.3:1, is shifted to 1:1 and is not infrequently 0.9:1. Under these circumstances, the additional dietary load of phosphate results in hyperphosphatemia and seizures. Since prepared infant formulas which contain less phosphate (500 mg/L) have essentially replaced formulas made from evaporated milk, less neonatal tetany is seen in the United States. In England, however, neonatal tetany is still observed, reflecting the use of formulas containing higher phosphate loads.[16]

In 1950, Harris and Nizel of Massachusetts Institute of Technology noted that hamsters fed a diet of corn and milk produced in Texas had fewer dental caries than hamsters fed a similar diet containing corn and milk produced in New England. Since the flouride content of each diet had been made equal, these investigators initially concluded that the difference in incidence of dental caries seen with the two diets was due to the presence of a cariogenic factor in the New England food. By doubling the ash content of the New England diet, dental caries were reduced. Further analysis of the ash indicated that mono-potassium phosphate was the active cariostatic agent.[17] Harris and his co-workers in subsequent years continued to investigate the role of phosphate as a cariostatic substance. They have concluded that the cariostatic effectiveness of various phosphates is anion, cation, and diet dependent.[18] The effect of anions is shown in Table 13. Concentrations of sodium phosphate in the diet of 0.8 to 1.4 percent are effective in reducing dental caries in rodents.

While the mechanism of phosphate action in caries prevention is not agreed upon by all investigators working in this field, prevailing evidence points to a local mechanism of phosphate in inhibiting caries. The effectiveness of phosphate appears to be dependent upon the period of time that the phosphate is retained in solution in plaque saliva. Phosphates administered by stomach tube or swallowed as pills are not cariostatic. Since adhesiveness or prolonged contact to the tooth surface with phosphate is important in caries protection, a choice of a foodstuff ingested by man that might provide a route of phosphate administration becomes important.

Several clinical studies have been carried out on children fed a variety of foods enriched with different forms of phosphate. In Sweden, Stralfors added calcium acid phosphate at a level of 2 percent to flour used in the school lunch program.[19] He reported an inhibition of caries among children ingesting this phosphate-enriched flour.

Ship and Mickelsen carried out a three-year study on calcium acid phosphate enriched flour in a series of schools in North and South Dakota and failed to find differences between placebo and experimental groups in decayed, missing, or filled teeth, or decayed, missing, and filled surfaces.[20] Averill and Bibby reported similar results from a study in New York State and a group of children studied in Brazil.[21]

Stookey, Carroll, and Muhler have reported the results of a study of 500 children between 5 and 16 years of age in Bloomington, Ind.[22] One group of children received pre-sweetened, ready-to-eat breakfast cereals containing 1.0 percent sodium diacid phosphate; the other group of children received similar cereal without the acid phosphate. Each child was examined at intervals of 6 months by 2 independent examiners. The results summarized in Table 14 indicate that the children receiving phosphate-enriched cereals showed a statistically significant lower DMFS rate than the children who received only the pre-sweetened cereal. The encouraging aspect of this study was the lowering of caries lesions on the proximal surface of teeth.

iodine Data released by the Salt Institute some months ago show the percent of use of iodized salt by geographic area of the nation.[23] In the Eastern part of the United States, about 52 percent of the salt used is iodized, in contrast with the Western part where some 58 percent is iodized, or the North Central area where use of iodized salt ranges as high as 61 percent. What we see in the preschool nutritional survey is that children in Northeastern United States excrete less than 100 micrograms of iodine per gram of creatinine.[24] The fact that urinary excretion of iodine appears to increase in areas with a reported higher use of iodized salt may indicate that iodized salt is more of a factor of daily iodine intake than bread containing iodate as a dough conditioner.

zinc and cadmium Widdowson and co-workers reported some years ago that the breast-fed infant in the first 10 days of life was in negative iron, copper, zinc, and manganese balance.[25] She postulated that during infancy some iron, copper, and manganese may be excreted in the stools because the infant, unlike the adult, lacks the ability to split these metals from the metallo-complex secreted in bile. Failure to resorb these metals produces negative balance. Whether zinc loss occurs via the same mechanism is not know.

We have carried out a large number of zinc balance studies on infants through 120 days of age (Table 15). At an intake of less than 0.7 gm per kilogram body weight, negative zinc balance was observed in infants 8 to 30 days old. When zinc intake approximated 1 mg per kg per day, positive zinc balance was obtained. Between the third and fourth months of life, positive balance was obtained at 0.75 mg per kg per day. Widdowson further reported that breast-fed infants were in positive cadmium balance.

Since cadmium and zinc react antagonistically, one might speculate that the strongly positive cadmium balance of the infant may contribute to its negative zinc balance. These observations would suggest that, in the fortification of foods with zinc, attention be paid to the cadmium content of the foods in question.

Table 10.
Vitamin D—Average Daily Intake and Contributing Sources.[8]

Age of Subject	Total	Source		
(yrs)	(IU)	Vitamins (IU)	Milk (IU)	Foods* (IU)
0-1	462	201	257	4
2-5	660	283	309	68
6-8	532	146	280	106
9-11	578	151	276	151
12-14	579	92	308	179
15-17	477	0	367	110
All	547	145	300	102

*Other than milk.

Table 11.
Vitamin K Content of Various Infant Feedings.[13]

Feeding	Vitamin K μg/L
Cow Milk	60
Human Milk	15
Sobee	80
Mullsoy	71
Isomil*	17-36
Enfamil	40
Similac	35
Nutramigen*	18
Meat Base	7-16

*Now fortified with vitamin K_1

Table 12.
Hypoprothrombinemia Associated with Infantile Diarrhea.[14]

Feeding	No. Infants*	No. Hypo-prothrombinemia
Cow's Milk	21	0
Casein Hydro-lysate	22	8

*All infants received Neomycin 100 mg/kg and IV fluids.

Table 13.
Cariostatic Activity of Various Phosphate Anions.[18]

Anion	Caries Score	Caries Reduc. %
Control	42	—
Ortho	28	33
Pyro	29	29
Tripoly	23	46
Trimeta	9	78
Hexameta	20	51

Table 14.
Dental Caries Increments in Children Fed Cereals Containing Phosphate.[22]

Time Mo.	Group	N	DMFT Incre.	DMFS Incre.	Prox. Surf. Incre.
			EXAMINER 1		
12	Control	100	1.95	4.85	2.24
	Phosphate	69	1.32[a]	2.81[c]	0.44[c]
24	Control	100	4.44	11.36	6.41
	Phosphate	69	2.77[c]	6.49[c]	2.87[c]
			EXAMINER 2		
12	Control	102	1.57	3.95	2.17
	Phosphate	67	1.40	2.93	0.74[c]
24	Control	102	3.92	9.35	5.14
	Phosphate	67	3.13	7.10[a]	2.98[b]

$P = (a) < 0.05; (b) < 0.01; (c) < 0.001.$

Table 15.
Intake and Retention of Zinc.

Formula	Age (days) 8-30		Age (days) 91-120	
	Intake mg/kg	Retention mg/kg	Intake mg/kg	Retention mg/kg
A	0.95	0.24	0.75	0.35
B	0.44	−0.09	0.40	−0.13
C	0.21	−2.00	0.16	−0.88
D	0.35	−0.50	0.30	−0.04
E	0.72	−0.26	0.56	−0.60

1. Infants and children are more sensitive to excessive intakes or imbalances of vitamins and minerals than adult man.

2. Preformed vitamin A has a relatively narrow margin of safety. Thus, its addition to foods should be controlled. Current practices of fortification with Vitamin A do not appear to constitute a hazard. There is little or no evidence to support a need for increasing the level of vitamin A added to food or to extend the addition of vitamin A to heretofore unfortified foods.

3. Confusion exists in food tables regarding the vitamin A equivalency of naturally occurring carotenoids. The food industry should make a more realistic appraisal of labeling in this respect.

4. The margin of safety of vitamin D is such that foods other than milk and infant formula products should not be enriched with vitamin D. A safe and effective intake for normal infants, children, and pregnant women is 400 International Units per day.

This level of intake appears readily attainable if this practice of enrichment is followed.

5. Infants on diets low in vitamin K have been observed to develop hypoprothrombinemia. If vitamin K is added to the diet, a normal prothrombin time is found. These observations relate to formulas whose protein or nitrogen source is not milk. In considering formulated foods of the future, enrichment with vitamin K may be necessary.

6. Phosphates may be added to evaporated milk, and formulas made from evaporated milk containing added phosphates when fed to infants will produce neonatal tetany.

7. Phosphates in cereal products have been shown to be cariostatic.

8. In considering trace element additions, interactions of ions such as zinc and cadmium may have an adverse action on zinc availability or utilization.

1. Statement of Committee on Drugs and Committee on Nutrition: "American Academy of Pediatrics, Use and Abuse of Vitamin A." *Pediatrics.* 48: 655, 1971.

2. Report of a Joint FAO/WHO Expert Group: "Requirements of Vitamin A, Thiamine, Riboflavin, and Niacin." *WHO Technical Report Series 362,* Rome, 1967.

3. Owen, G. M., Garry, P. J., Lubin, A. H., and Kram, K. M.: "Nutritional Status of Preschool Children: Plasma Vitamin A." *J. of Pediatrics,* 78: 1042, 1971.

4. Underwood, B. A., Siegel, H., Weisell, R. C., and Dolinski, M.: "Liver Stores of Vitamin A in a Normal Population Dying Suddenly or Rapidly from Unnatural Causes in New York City." *Am. J. of Cl. Nutrition,* 23: 1037, 1970.

5. Statement of Committee on Nutrition, American Academy of Pediatrics: "The Prophylactic Requirement and the Toxicity of Vitamin D." *Pediatrics,* 31: 512, 1963.

6. Statement of Committee on Nutrition, American Academy of Pediatrics: "Vitamin D Intake and the Hypercalcemic Syndrome." *Pediatrics,* 35: 1022, 1965.

7. Statement of Committee on Nutrition, American Academy of Pediatrics: "The Relation between Infantile Hypercalcemia and Vitamin D—Public Health Implications in North America." *Pediatrics,* 40: 1050, 1967.

8. Pale, A. E. and Lowenberg, M. E.: "Consumption of Vitamin D in Fortified and Natural Foods and in Vitamin Preparations." *J. of Pediatrics,* 70: 952, 1967.

9. Seelig, M. S.: "Vitamin D and Cardiovascular, Renal, and Brain Damage in Infancy and Childhood." *Ann. of N.Y. Acad. of Sciences,* 147: 537, 1969.

10. Goodenday, L. S. and Gordan, G. S.: "Fetal Safety of Vitamin During Pregnancy." *Clinical Research,* 19: 200, 1971.

11. Rapoport, S. and Dodd, K.: "Hypoprothrombinemia in Infants with Diarrhea." *Amer. J. Dis. Child,* 71: 611, 1946.

12. Goldman, H. I. and Desposito, F.: "Hypoprothrombinemia Bleeding in Young Infants." *Amer. J. Dis. Child,* 111: 430, 1966.

13. Goldman, H. I. and Amadio, P.: "Vitamin K Deficiency After the Newborn Period." *Pediatrics,* 44: 745, 1969.

14. Goldman, H. I., Desposito, F. and Sawitsky, A.: "Diet, Diarrhea, and Vitamin K Deficiency Bleeding." The Society for Pediatric Research, Thirty-Seventh Annual Meeting, 99, April 28-29, 1967.

15. Statement of Committee on Nutrition, American Academy of Pediatrics: "Vitamin K Supplementation for Infants Receiving Milk Substitute Infant Formulas and for Those with Fat Malabsorption." *Pediatrics,* 48: 483, 1971.

16. Oppe, T. E. and Redstone, D.: "Calcium and Phosphorus Levels in Healthy Newborn Infants Given Various Types of Milk." *Lancet* 1: 1045, 1968.

17. Nizel, A. E. and Harris, R. S.: "The Effects of Phosphates on Experimental Dental Caries: A Literature Review." *J. Dent. Res.,* 43: 1123, 1964.

18. Harris, R. S., Nizel, A. E., and Walsh, N. B.: "The Effect of Phosphate Structure on Dental Caries Development in Rats." *J. Dent. Res.,* 46: 290, 1967.

19. Stralfors, A.: "Effect of Calcium Phosphate on Dental Caries in School Children." *J. Dent. Res.,* 43: 1137, 1964.

20. Ship, I. I., and Michelsen, O.: "Effects of Calcium Acid Phosphate on Dental Caries in Children: A Controlled Clinical Trial." *J. Dent. Res.,* 43: 1144, 1964.

21. Averill, H. M., and Bibby, B. G.: "A Clinical Test of Additions of Phosphate to the Diet of Children: A Controlled Clinical Trial." *J. Dent. Res.,* 43: 1150, 1964.

22. Stookey, G. K., Carroll, R. A., and Muhler, J. C.: "The Clinical Effectiveness of Phosphate-Enriched Breakfast Cereals on the Incidence of Dental Caries in Children: Results after 2 years." *JADA,* 74: 752, 1967.

23. Report of the Salt Institute, 1970.

24. Owen, G. M., Kram, K. M., Garry, P. J., Lowe, J. E., Jr., and Lubin, A. H.: "A Study of Nutritional Status of Preschool Children in the United States, 1968-1970." To be published 1973.

25. Widdowson, E. M.: "Trace Elements in Human Development in Mineral Metabolism in Pediatrics." Ed. by Baltrop, Donald and Barland, W. L.: F. A. Davis Co., 85, 1969.

Q: We'd like to ask if there's any evidence at all that a small supplement of chromium in Cr-deficient persons may have an effect in reducing serum-cholesterol level? We'd heard rumors to this effect.

A (Dr. Mertz):   This is a very difficult question to answer. Schroeder has shown that under carefully controlled conditions in rats, chromium does indeed lower cholesterol levels. We have been trying to do it and under laboratory conditions we have not been able to significantly influence cholesterol. The difference may be in the degree of chromium deficiency, which was more pronounced in Shroeder's animals than in ours.

Q: Many years ago we had lots of animal toxicity from selenium in grain and the Public Health Service at that time looked into the question of human toxicity. They found people with high-level urinary excretions of selenium but, as we remember it, no demonstrable toxic effects. Would you tell us the current situation?

A (Dr. Mertz):   We are aware of only one study on possible human toxicity of selenium, in the teeth of children living in high selenium areas. Some enamel pathology caused an increased caries susceptibility in the children living in the high selenium area. This, as far as we know, is the only instance of selenium toxicity.

Q: Is there any biochemical parameter which could be used to detect selenium influence, short of usual pathology?

A (Dr. Mertz):   You could probably use selenium excretions, but we know so little about the biochemistry of selenium that we cannot think of an easy test.

Q: In relation to the discussion of vitamin D and vitamin D toxicity, we bring to your attention a paper that was published from the Toronto Hospital for Sick children on iatrogenic rickets in low birth weight infants. Have you seen this?

A (Dr. Filer):   Yes.

Q: Here is a case where, using the good practice of having formula with 400 units of vitamin D per quart, the hospital decided some time back to stop using vitamin supplements because they felt that the formulas gave all of the vitamins and minerals that were needed. Then, these low birth weight infants came along and, because of their low intake of formula, developed rickets. Therefore, it is perfectly fine to keep a balance between levels of intake and levels at which we fortify with nutrients, but in special situations we have to concern ourselves about total intake.

A (Dr. Filer):   This Canadian practice is not much different from what we do in the United States, and we really haven't seen this kind of syndrome. The Canadians spot a lot of things that we don't see, and do some things we don't do. We are not sure that we understand what this recent Canadian report means.

Q: The data cited from the Salt Institute shows, in comparison with the iodine excretion rate, a high usage of iodized salt in the North Central and the Atlantic States. Chapter 3 did not comment on this. Does it imply that the iodine in salt is not utilized, that it's too low or that there are other factors involved?

A (Dr. Filer):   About 51 percent use iodized salt in the Northeast-North Central States in contrast to 61 percent in the Midcentral-West Central States, and the excretions follow that pattern. This says that a distribution pattern of the use of iodized salt bears a close resemblance to urinary excretion patterns. We were impressed more with the similarity in the two separate data than we were with the dissimilarities.

Q: We were impressed that there is greater regional difference in excretion data than in the Salt Institute's estimates of the use of iodized salt. One can get up to 100-fold of the physiologic level of iodine from dough conditioners where iodate is used as a dough conditioner in the continuous mix bread process. Interpretation of data on iodine excretion must take into account the varying use—region to region and city to city—of this process and of the use of iodate.

A (Dr. Cotton):   Several years ago when we observed that iodate in continuous mixed bread interfered with certain medical diagnostic procedures, we removed the iodate from use. We wouldn't be surprised that other bread people have, also.

Q: Was there then a substitute bromate for the conditioner? The question was raised in the Food Protection Committee's Iodine Workshop as to whether there is any evidence of an interaction between bromate and iodine or bromine and iodine in metabolism, whereby the presence of bromine might increase or alter the effectiveness or the metabolism of iodine or the requirement of iodine?

A (Dr. Cotton): We don't offhand know the levels used, but the use of bromate is very much lower than the use of iodate in continuous mix bread.

Q: Is there any danger of hypervitaminosis A from feeding carrots to infants or is there some self-limiting factor in the conversion that would make high levels of carotene less likely to produce hypervitaminosis A?

A (Dr. Filer): The point Chapter 3 was trying to make was that, in labeling, one should not oversell the vitamin A content of carrots or sweet potatoes. Use a more realistic conversion figure and put a reasonable figure on the jar.

Q: Two publications in Great Britain on food faddists report on adults who consumed between 1-1/2 and 2 kilograms of carrot juice and tomato juice a day. There was no evidence in either person of hypervitaminosis A.

A (Dr. Hein): We know of nothing bad on carotene. In the case of tomatoes, pigmentation of the skin may not be due to a pro-vitamin A (carotenoid); it may be due to lycopene. In addition, carotene is excreted and, with very high amounts of carotene being administered, very large amounts would be excreted in the feces.

Q: Chapter 3 mentioned that there was no need for any further vitamin A fortification, for infants and children. How would you rationalize that opinion with the data that Chapter 1 presented showing a rather sizable portion of children of all ages having low vitamin A levels? Also, Chapter 3 mentioned that milk was the single best source of vitamin A for children, and obviously it referred to whole milk. With the increased use of skimmed milk and the low-fat milk, do you think it is desirable that they be fortified

with vitamin A up to the level of whole milk? In the not too distant future, skimmed milk and low-fat milk will probably be used in the school lunch programs and, if so, should they be fortified to whole milk levels?

A (Dr. Filer): We don't think Chapter 1 showed us any specific age group other than the 10- to 12-year group. The other data pertained to the entire population. We think Chapter 1 showed us 2.5 percent incidence of vitamin A levels, less than 20 mg per 100 gm. We think that a 4 percent figure is realistic for the nation as a whole. We think you misunderstood Chapter 3's reflection of the distribution of the vitamin in milk. This happened to be vitamin D and not vitamin A. It didn't show any data on the contribution of milk to our total vitamin A intake. But we agree that skimmed milk, either powdered or liquid, should be fortified with vitamins A and D, to the level found in whole milk.

Q: In the 6-year-old category by states, we'll read the percent that fell below 20 mg per 100 gm, and these are, respectively—we don't mention the states—18, 25, 22, 23, 13, 40, and 5—and in two states there were none. We could go a year before that—the 5-year-old group—15, 31, 10, 19, 0, 17, and 10—and two other states with none. The 4-year-old group is 50, 26, 27, 30, 18, and 5 percent—and three states with none. So we do have a sizable percentage of children under 6 as well as under 17 that have levels of serum vitamin A below 20 mg.

A (Dr. Stare): The issue is, how does one get at these people? We don't think it involves putting more vitamin A in foods or extending the foods that are fortified with vitamin A. We think we need to feed these people the food that we eat, foods that presumably carry sufficient vitamins.

Q: Chapter 3 says that vitamin D should not be added to anything other than infant formulas and milk, and we ask what we do about the lactose-intolerant children? If the work of George Graham at Hopkins means anything, lactose intolerance is a very significant thing in school children. Also, we have a hard time rectifying that with the data that Chapter 1 has presented in terms of vitamin D and rickets. In regard to iodine, the major relationship between the consumption of iodized salt and the amount of iodine that shows up in the urine doesn't relate to the incidence and distribution of goiter. There's something wishy-washy about that particular situation.

A (Dr. Filer): It would be nice to have the opportunity to review Dr. Graham's data on lactose intolerance. If you judiciously feed the African skimmed milk and he isn't a malnourished Biafran, he tolerates the skimmed milk rather well. We are trying to sort out what is myth and what is fact about this lactose intolerance thing. Within the last five or six years this became a very popular clinical diagnosis, and we suppose we gave more lectures on disaccharide intolerance than we should to general practitioners and medical students. We remain to be convinced that this is a problem.

With respect to vitamin D and the bowed legs and the rachitic rosary described in the survey, we don't think that these kinds of observations are always supported with alkaline phosphatase data or x-ray evidence of rickets. That child was, maybe, four or five year of age and we may have been seeing the ravages of earlier rickets. On the iodine question, you're calling an enlarged thyroid a goiter, and we think this is a common mistake. Chapter 1 very carefully pointed out that he was talking about enlarged thyroids rather than goiter. We don't think that we've totally handled the problem of natural goitrogens in the diet and they may be a factor in this problem.

Q: When we measure acceptance of food items in the New Jersey school system, we find children who eat other things than milk. So, where are they going to get calcium and vitamin D? Are you going to give them shots? We can't agree with saying we've got to measure forever to define the problem, rather than going out and trying solutions and trying to measure the impact of what we're doing at the same time.

Q: It was suggested that we have functional tests by which we may evaluate marginal states of micronutrient needs. Now, the question is, should micronutrient intake be set at such a level which will support maximal biochemical or physiological function for which that micronutrient is needed?

A (Dr. Mertz): If there is a wide enough margin of safety, yes, intake should be set for a near maximal function. It should be set to obtain a function which just starts approaching the flat part of the curve. If we ever have to think about selenium enrichment, for example, we will have to be very careful and allow a wide enough margin of safety.

factors influencing vitamin-mineral content of foods and biological availability

PART TWO

INTRODUCTION TO PART II
by C. O. Chichester, PhD

rapid growing ↓ nat... extra attn
α effect additives

A number of factors which influence the availability of vitamins and minerals to the human population arise from the production of the food material itself. We have paid relatively little attention historically to developing food on the basis of vitamin and mineral content. We are just now beginning to worry about the genetic problem of protein content of agricultural products. In most cases, we have bred largely for total yield per acre, or adaptability, or disease resistance.

In processing, which I think we must consider as one of the more important aspects of food in the United States, we have paid attention to the vitamin and mineral content of food. Yet, when we discuss what the family utilizes as its food source, many of us do not realize the extent of the use of processed foods in the United States. Approximately 70 percent of the foods that we consume are processed in one manner or another. Obviously, of major importance in the micronutrient content of foods is the effect of processing.

Storage and distribution may give rise to nutritional problems simply because chemical compounds in foods react proportionally to the time and temperature at which they are held together. When we assay the biological availability of nutrients in food, we must consider the analytical problems. We pointed this out earlier in discussing lactose intolerance. Lactose intolerance is a relative thing. The method for determining lactose tolerance is to give a massive dose of lactose; and, if the subject cannot handle it, his problem is defined as lactose intolerance. I'll warrant that most adults given a massive dose of lactose would prove to be, under these conditions of analysis, relatively lactose intolerant.

In agricultural practices, I think there is evidence that some new varieties of food plants may differ from their predecessors in content of vitamins and minerals. An example is the yellow watermelon versus the traditional red watermelon. The red watermelon probably has no vitamin A activity. The yellow, on the other hand, is a good source of pro-vitamin A. It has been said that there is some evidence that the ascorbic acid content in presently harvested varieties of tomatoes is lower than in previous varieties, although this is debatable.

38

There is little evidence to show that agricultural practices introduced over the past three decades, including new high-yielding varieties, increased rates of fertilizer use, and the application of herbicides and soil fumigants have, with the possible exceptions of zinc and magnesium, resulted in a decrease in the vitamin or essential mineral content of food crops. The introduction of rapid methods of harvesting and packaging of vegetables and fruit and their transport in refrigerated equipment of improved design has probably contributed to delivery of these foods to the retail market with higher vitamin content. There is limited data, however, on the vitamin and mineral content of food crops as produced three or more decades ago on which a reliable comparison can be made with current production. Best data is available for protein content of wheat and corn and these indicate that relatively small change (± ½ percentage point) has accompanied increased yield of these crops.

Sufficient genetic variation for ascorbic acid and carotene content exists in current breeding stocks of some vegetables and fruits such that varieties could be developed with doubled or more content of these vitamins as compared with current commercial varieties. Such development would be desirable as a means of providing increased vitamins for consumers of both fresh and processed foods. In developing varieties for increased nutrient content, and particularly for other desired plant characteristics such as insect or disease resistance, the plant breeder needs to be aware of the possibility of introducing or increasing toxic constituents.

The need to increase the vitamin content of a number of food crops in order to improve the nutrient intake of our population which has diverse dietary habits is illustrated by examination of the food selections of people over 65 years of age. In view of the large and increasing proportion of food processed, and the large losses of nutrients that may occur in handling and processing, it is proposed that restoration of vitamins and essential minerals in the processing operation would be a feasible and economical method of improving the nutrient content of a variety of foods.

The national average yields of some of our most important crops, such as wheat and corn, have more than doubled during

chapter 4
agricultural practices influencing vitamin-mineral content of foods and biological availability

by
Frederic R. Senti, PhD

the last two or three decades. Today's high yields are possible because of the application of a group of farm practices, including new high-yielding varieties of crops and increased rates of fertilizer use. This upsurge of productivity has led to an unprecedented abundance and variety of food in the developed countries, and under the label of "The Green Revolution," it is helping to dispel the threat of hunger from the developing nations. At this time, I want to consider the question of whether or not this abundance and variety of foods has been purchased at the cost of a decline in the nutritional quality of this food supply.

I shall also discuss briefly the occurrence of toxic plant constituents, some of which may increase the requirement for essential minerals in human nutrition, and which may be intentionally or inadvertently changed in varietal selections. Finally, the problem of selection of a food crop or product for varietal improvement in nutrient content in order to reach a maximum number of those deficient in that nutrient will be illustrated by considering the food habits and dietary intake of vitamin C by people in the population group over 65 years of age.

The effect of recently introduced agricultural practices on quality of our food supply is a controversial topic. The claims of some of the "food fad" exponents have been widely publicized. Recent nutritional surveys have pointed to important instances of malnutrition among certain parts of the population. And yet, our national public health statistics continue to show decreasing percentages of infant deaths and increasing length of life expectancy, and the incidence of nutritional diseases such as scurvy and rickets is low. We feel that much of the controversy about food quality can be traced to excessive generalizations—and that we can have meaningful discussions of these problems only if we consider specific required nutrients, specific diets, and specific systems of food production.

MINERAL ELEMENTS

Humans and animals require many different nutrients for optimum health. Plants serve as primary sources of many essential nutrients, and a typical pathway of essential minerals into human diets involves transfer from the soil, to the plant, to man, with animal food products being a very important intermediate step, in some cases. Our knowledge of dietary requirements is being steadily improved.

During the past 12 years, selenium and chromium have been added to the list of elements required by animals (and probably by man) and recent evidence indicates that nickel, vanadium, and tin may soon be confirmed as essential. A healthy, high-yielding plant does not necessarily contain enough of some of the essential minerals such as zinc, chromium, iodine, selenium, iron, manganese, and others to meet all of the requirements of the person or animal that eats this plant. We need to consider each essential mineral separately.

Turning first to iron—it is apparent that a marginal anemia due to iron deficiency is prevalent in women of childbearing age in most parts of the world, including the United States. In all likelihood, this is not a new problem, and food crops raised at today's high-yield levels are no worse than the low-yielding crops of former times as far as their value as sources of dietary iron is concerned. Nearly all soils contain large amounts of iron, but the ability of the plant to take up this iron is affected by many factors, including the acidity of the soil and the species or variety of plant.

Plants suffering restricted growth due tc iron deficiency frequently contain almost as much iron per unit of weight of plant as do iron-adequate plants. There is evidence that people do not utilize the iron contained in vegetables or cereals as well as they do the iron in meats, and so an increase of the plant available level of iron in agricultural soils would not necessarily correct iron deficiency in people. Direct addition of supplementary iron to packaged food products is a current practice, and the level of iron supplementation recommended is being increased.

The need for zinc by both plants and animals has been recognized for a long time, and there is growing evidence that significant zinc deficiency occurs among residents of the United States. A zinc-deficient person does not heal as rapidly following surgery or wounds as a person with adequate zinc supply. On the basis of increasing reports of needs for zinc fertilization of crops, we feel that it is quite probable that the level of zinc in food plants may be declining. In addition, some other sources of zinc for people, such as galvanized plumbing and utensils, are being replaced by other materials. So zinc deficiency is a problem for both humans and animals.

Attempts to meet this problem by increasing the level of plant available zinc in farm soils are complicated by the fact that many crops will make optimum yields, even though they contain insufficient zinc to meet requirements of the animal that eats the crop. The zinc contained in plants is only partially utilized by animals and, presumably, by people. The zinc in foods of animal origin, such as meat and eggs, is utilized by monogastric animals better than zinc from plant foods. Thus, food selection is likely to be an important factor in the zinc nutrition of people, and the high agricultural productivity that has provided our current abundance of meat and eggs may have helped to prevent widespread critical deficiencies of zinc.

Chromium is the most recent addition to the list of mineral elements known to be required by man. Chromium acts in the body as an activator of insulin, and chromium deficiency leads to abnormal metabolism of sugars and to diabetes. Some diabetic patients have responded to chromium plus insulin treatment after response to insulin alone had been disappointing. There is no way of estimating whether or not diets in the United States contain less chromium now than in former years.

The discovery of an important role for chromium is very recent, and methods for detecting and measuring the small amounts of chromium involved in this role are based on even more recent work. In our Human Nutrition Research Division, we are pursuing efforts toward understanding the role of chromium in nutrition, and the identification of the chemical forms of chromium that are effective dietary sources of this element. In the Soil and Water Research Division, some of our people are working on the processes involved in the soil-to-plant-to-animal movements of chromium, and studying techniques for increasing the levels of chromium in food plants.

At one time, the element selenium was of interest to nutritionists only because of its toxicity to animals and man. It was implicated as a carcinogen, and the addition of selenium to diets for animals and humans was prohibited. Then about 12 years ago, selenium was identified as an essential factor for animals, and it was found that some very important nutritional diseases of farm animals could be prevented by addition of very low levels of selenium to their diets.

Agricultural Research Service (ARS) scientists then prepared a map of the United States in which areas where the crops contained too little selenium to meet animals' needs were distinguished from the areas where plants contain adequate, but nontoxic, levels of this element (Fig 7). As the map took form, it became evident that for many years we have been modifying the levels of selenium in the diets of animals and people in some parts of the United States by shipping grains from the adequate areas to regions of deficient selenium supply. The adequate levels of selenium in crops grown in the Plains States were reflected in higher levels of selenium in milk produced there. Even in human blood, a difference in selenium level could be detected between residents of the adequate and inadequate regions.

About this time, evidence that has led to questioning the earlier implication of selenium as a carcinogen began to appear. There is some evidence that selenium is required by people, and that some diets may be deficient in this element. The question of what to do about selenium in human and animal feeds is a difficult one, because even though its carcinogenic implications have been challenged, it is well known that some chemical compounds of selenium are very toxic—perhaps more toxic than equal levels of arsenic. Decisions as to whether or not to permit addition of selenium to animal feeds are currently being given serious consideration by the appropriate agencies of government.

In many cases, it is very difficult to determine whether or not the new agricultural practices for securing high yields have led to changes in the concentration of essential minerals in human foods. This is primarily due to the fact that we have relatively little information on the concentration of these minerals in the foods that were available some years ago. We can estimate, on the basis of our knowledge of soils and plant nutrition, that the concentration of zinc and magnesium in food crops has probably declined over the past 30 years. There may be other elements in this same group, but we also know that some mineral deficiencies have been prevalent for a very long time.

Shakespeare mentioned the prevalence of goiter among mountaineers. The discovery of the relation of iodine deficiency to goiter incidence, plus the correction of iodine deficiency through the use of iodized salt, is one of the classic contributions of science to human welfare. A deficiency of cobalt for cattle and sheep in parts of the eastern United States was a serious problem during the colonial period. Iron deficiency anemia is undoubtedly a problem of long standing. Although the history of some of these problems is very interesting, our primary concern needs to be with the problems of the present and future quality of the food supply.

In addition to the mineral elements that are essential to human life, we must be concerned about trends in the concentrations of lead, cadmium, mercury, and other elements from agricultural crops which have not yet been clearly established. This is because their innate toxicity requires that we know the concentrations of these elements in our foods, and know whether or not these concentrations are dangerously increasing.

We may be able to design agricultural practices that will minimize the levels of some of the toxic elements in foods by inactivating these elements in the soil so that they will not be taken up by plants. In other cases, it may be possible to remove toxic elements during food processing, without otherwise damaging the quality of the foods. Our research program includes efforts along both of these lines.

PROTEINS—QUALITY AND QUANTITY

Protein malnutrition, due to a low amount of protein or to low-quality protein in human diets, is a major problem for residents of the developing countries and for certain low-income groups in the United States. Will the new varieties of crops, grown at today's high rate of fertilizer use, help solve this problem, or will they make it worse? The field crops and vegetables that are being produced so abundantly today generally contain as much protein as the lower-yielding crops of former years, and most of the available evidence indicates that the nutritional quality of the protein has been maintained.

In the case of wheat, our principal food grain, average protein content has probably declined slightly. The longest continuous record on the protein level for a major wheat-producing area is for Kansas; this record is complete from 1949 to 1970. During the period 1949-58, the average in Kansas was 12.4 percent; in the period 1957-66 it was 11.8 percent. Shorter periods are available for Montana, North Dakota, and a few other states. It might be generalized that a loss of ½ percent protein has occurred in the hard red winter wheat belt.

Higher rates of nitrogen fertilization have largely resulted in higher yields of grain without changing protein content. For example, in the Pacific Northwest, Gaines, a semi-dwarf wheat, is fertilized at a 25 percent heavier rate than tall varieties, yet the protein content is the same. In the Ohio Valley, we often hear complaints from millers that the protein of soft wheat is higher than desired as a result of heavy fertilizer applications.

Data on the protein content of corn collected from the principal markets in the corn belt over the period 1943-57 indicate a slight increase in protein level. Average protein content of the period 1943-47 was 9.69 percent; that for the period 1953-57 was 10.18 percent—an increase of about ½ percent. For these same two periods, average yield of corn in Illinois, one of the major corn-producing states, increased from 46 to 58 bushels per acre.

Breeding for higher protein in wheat holds promise of raising the level 1 to 3 percentage points without a lower yield. Pioneering work in North Carolina—followed by work in Nebraska, Indiana, and Kansas—confirms and extends this conclusion. Advanced generation lines with higher protein content are available for use in breeding programs.

There also is promise that protein quality in cereal grains can be improved through breeding. An outstanding advance was made in the discovery that certain mutant lines of corn had relatively high contents of both lysine and tryptophan—the essential amino acids in which ordinary corn is deficient. High lysine inbred and hybrids are being developed by both public and private agencies. None of the hybrids so far developed have given satisfactory yield levels and it may be some years before extensive commercial use is made of such materials. Encouraged by the success in developing corn hybrids that have protein of improved nutritional quality, we are now working on the development of varieties of wheat, oats, and other crops that will have genetic inheritance to form proteins richer in the amino acids needed by man.

YIELD INCREASES AND THE VITAMIN CONTENT OF FOODS

Plants serve as primary sources of several important vitamins in human diets. Some of the food plants are important dietary sources of one or more vitamins, and of minor importance as sources of other vitamins. Citrus fruits and tomatoes are very important sources of vitamin C (ascorbic acid), and whole wheat products may make appreciable contributions of iron, niacin, and thiamine to humans.

There has been recent publicity over the value of vitamin C in human diets. The factors affecting the vitamin C level in tomatoes were the subject of intensive investigations in ARS some time ago. Some of the important findings are illustrated in Fig 8. The amount of sunlight received by the ripening fruits is the major factor controlling the concentration of vitamin C in tomatoes. The use of fertilizers has relatively little effect on the concentration of this vitamin in tomatoes, unless the fertilizer caused a luxurious growth of foliage that shaded the ripening fruits.

Commercial tomato growers try to avoid the use of fertilizer applications that would cause excessive foliage and shading of the fruits, because they want their entire crop to ripen over a short period. This short ripening period helps to cut labor costs at harvest. So—it appears that the current production practices used for tomatoes have not brought about any marked reduction in the vitamin C content of the marketed crop. This is probably also true for other vegetables and for citrus fruits.

Humans and animals also depend upon plants as a primary source of carotene (pro-vitamin A). Investigations of factors affecting the carotene concentration in plants have shown that fertilizer treatments can result in increases in the carotene concentration in plants, providing the fertilizer treatment corrects a yellowed condition of the plant. This yellowed condition is usually termed "chlorosis." A yellowed condition of plants that should normally be green may be the result of deficiency of several of the many nutrient elements required by plants. No single fertilizer practice will automatically insure plants with high levels of carotene.

There are records of increased carotene in plants following fertilization with boron, iron, and other elements. The fertilizer practices used must be specifically fitted to the particular soil and crop. Since a yellowed condition of the leaves of plants usually is associated with low yields, modern farmers have attempted to correct this condition as rapidly as possible. It appears to be quite likely that the high-yielding crops of the present time are at least as good, and probably better,

sources of carotene than those available some time ago.

VITAMINS IN FRUITS AND VEGETABLES

Vitamin content of vegetables and fruits may differ depending on variety, and varietal selection has led to increased content of vitamins in several instances. Data on the range in carotene and ascorbic acid content found in commercial and in breeding lines of some vegetables are presented in Table 16.

Orange carrot roots are excellent sources of carotenoids, especially alpha and beta carotenes. Color of the carrot root in various breeding lines ranges from white (traces of carotenoids) to dark orange containing up to 370 mg of total carotenoids per 100 gm of fresh weight. There is a positive correlation between color intensity and total carotenoids; therefore, highly colored carrots are more desirable commercially, not only for their better appearance, but also for higher nutritive value (pro-vitamin A).

Breeding programs, therefore, have been directed toward higher nutritive value for some time. The inheritance of color (and carotenoids) has been worked out and is being put into use by some carrot breeders. The mean carotenoid content of 7 orange varieties was 64 to 94 mg/100 gm fresh weight. The reported content of 370 mg/100 gm for a breeding line represents the potential in carrots.

Raw cabbages are a good source of ascorbic acid. Present commercial varieties average 50-60 mg/100 gm fresh weight. An advanced breeding line developed at the United States Breeding Laboratory at Charleston, S. C. averaged 93 mg/100 gm. This line was derived from a collard cross, the source of the high ascorbic acid content.

Muskmelons are a good potential source of ascorbic acid since they are eaten raw. Breeding lines at the US Vegetable Breeding Laboratory range from 3 to 61 mg/100 gm of edible flesh. Commerical varieties average around 30 mg. The most recent variety released from the US Vegetable Breeding Laboratory at Charleston, S. C., "Planters Jumbo," averaged 47 mg. There is a close positive correlation between ascorbic acid and sugar content in melons.

Tomato varieties that are high in

vitamin content have been released in this country. The variety, Doublerich, was developed by crossing the garden tomato with *Lycopersicum peruvianum* (a small-fruited, wild type) and selecting for large fruits and high vitamin C. Ordinary varieties have 15-25 mg of ascorbic acid per 100 gm fresh fruit. Doublerich, developed at the New Hampshire Agricultural Experiment Station, had more than 50 mg. There is no evidence, however, that Doublerich has been demanded by the home gardeners or commercial growers for its source of vitamin C.

Caro-Red, a variety developed at Purdue University, produces fruit containing approximately 10 times the pro-vitamin A content of common red tomatoes. The beta-carotene values for Caro-Red are within the lower ranges reported for carrots. These two varieties are not widely grown, but are good sources of germ plasm for future interest in developing high vitamin tomatoes.

Sweet potatoes are a rich source of vitamin A. Like carrots, the deeper colored flesh indicates a higher carotene content and hence a higher level of vitamin A. In recent tests, carotene pigments in a group of about 300 new seedlings ranged from 5 mg/100 gm to 22 mg/100 gm. This range indicates the very high range of variability that exists within a population of sweet potato seedlings and indicates that breeding for extremely high carotene content is feasible.

Standard varieties on the US market in the same test had pigment contents of 5 to 12 mg/100 gm, the highest being Centennial, a leading market variety. The vitamin A value that we have used in evaluating the contribution of sweet potatoes to the US diet has increased over the years from around 7,700 IU per 100 gm in the 1940s to about 8,800 in the 1960s, and may continue to increase as new varieties having flesh of deep orange are developed and replace older varieties with light-colored flesh.

Fruits are recognized as being relatively rich sources of vitamins. Strawberries (breeding lines) have been selected for high vitamin C content. A Department PL-480 project on collecting and studying indigenous, wild, small-fruited types and breeding varieties with a high vitamin C content, is now being supported in Yugoslavia. There also appears to be a good potential in the currant, blueberry, and blackberry for breeding lines with high vitamin C content.

We can conclude that there is the necessary variability in fruits and vegetables to increase their content of ascorbic acid and carotene and possibly other vitamins through genetic development. In some cases, increase has already occurred as a result of selection for characteristics, eg, color associated with higher vitamin content. In other cases, high vitamin content has been the primary objective in the development of a new variety.

COBALT-DEFICIENT SOILS IN RELATION TO VITAMIN B_{12}

Vitamin B_{12} gets into human diets through a complicated pathway. This vitamin contains the trace mineral, cobalt. Plants take up cobalt from the soil, but higher plants do not synthesize the complete molecule of vitamin B_{12}. The major source of this vitamin is from the microorganisms in the rumen of cattle and sheep. These microorganisms utilize the cobalt contained in the plants eaten by the animal and synthesize vitamin B_{12}, which is then transferred to human diets through meat and milk.

Following the discovery by Australian scientists of the essentiality of cobalt for ruminants, research workers in the US Department of Agriculture (USDA) and several state experiment stations established that forages grown in certain areas of the eastern United States contained too little cobalt to meet ruminant animal requirements. The areas where clover and alfalfa are deficient in cobalt were located through field sampling and analysis of plants for cobalt. As a result of these studies, farm animals in the United States are frequently supplemented with cobalt—sometimes in the salt mix and in a few cases by adding cobalt to pasture soils. The recognition and correction of cobalt deficiency in animals, plus the development of a highly productive animal agriculture in the United States, has undoubtedly improved the supply of vitamin B_{12} for the people of this country.

VITAMINS AND MINERALS IN CEREAL GRAINS

Cereal grains, particularly wheat, make important contributions to the iron, thiamine, niacin, and riboflavin in the US diet. Although whole grain products contain significant amounts of these essential nutrients, as well as calcium, magnesium, and trace minerals, up to 70 to 80 percent loss occurs in milling to produce white flour, the most used wheat product. It was in recognition of this loss of vitamins and minerals that occurs in the milling, not only of wheat but also in corn and rice, and the important role of cereals in the US diet, that enrichment of cereal products with niacin, thiamine, riboflavin, and iron was initiated in the early 1940s. As far as these nutrients are concerned, the effect of fertilization—yield and variety—is not of importance to enriched cereal products.

It is known, however, that thiamine content, for example, is a heritable characteristic and that, if desired, thiamine content of wheat could be altered by breeding. Possibly, changes in the content of some vitamins and minerals in wheat may have occurred with introduction of new varieties, fertilization, and other agricultural practices. However, the limited compositional data available on these nutrients for crops raised two or three decades ago, together with the marked influence of climatic, soil, and environmental factors on composition, make it difficult to do any reliable comparisons.

EFFECT OF HERBICIDES AND SOIL FUMIGANTS ON NUTRIENT COMPOSITION OF PLANTS

Several investigations have been reported on the effect of application of herbicides on the protein content of cereal grains. Simazin, 2,4-D, dicamba, and other herbicides tend to increase protein content of cereal small grains, but protein elevation generally appears to be accompanied by loss in grain yield. The decrease in grain yield may lower the total protein production per acre. No change in protein composition has been reported to result from the herbicide treatments.

Some fungicides applied to the wheat plant are readily absorbed, certain ions are translocated and are deposited in the grain. Calcium sulfamate applied for rust control markedly impairs germination of the grain, as well as its baking value.

Carrots grown in soil treated with the soil fumigants Telone and Nemagon have been reported to be 17 to 74 percent higher in total carotene than were the control carrots. Sugar content of the carrots also was increased. In other experiments, carrots grown in soil treated with the herbicides CIPC and linuron were 20 to 25 percent higher in total carotene than were carrots grown in non-treated soil.

Butternut squash grown in soil treated with the herbicides amiben and dinoseb were 30 to 70 percent higher in carotene than control squash.

It thus appears that herbicides and soil fumigants may have a favorable effect on the vitamin content of some food crops. In the limited number of examples cited, no adverse effect on nutrient composition has been reported. However, when protein content was increased, a corresponding decrease in starch content usually occurred.

EFFECT OF HARVESTING, HANDLING, AND STORAGE PRACTICES ON VITAMIN CONTENT OF FOOD CROPS

Harvesting, handling, and storage practices for fresh agricultural food products have been revolutionized in the last three decades. Tractors replaced the horse as source of power on the farm and advances were made in mechanical harvesting stimulated by the steady decrease in farm labor supply and higher wages. With these changes it became possible to harvest crops faster and remove them from the field to the packing shed or processing plant with a minimum of exposure to unfavorable field conditions.

Also these changes made it possible to harvest large acreages when the crops were at the optimum stage of maturity and quality for processing or shipment to the fresh market. The rapid transportation to the packing shed and processing plants reduces exposure to high temperatures which cause undesirable physiological and compositional changes and possible loss of vitamins.

Mechanization in the packing shed has made it possible to grade, wash, precool, and package the products within a few hours of harvest and place them in cold storage or in refrigerated rail cars or trucks for shipment to market with little, if any, delay. Most lettuce and some other leafy vegetables are vacuum cooled shortly after harvest. This process reduces the temperature to the center of the packaged heads from about 85° F to 32° F in about 20 minutes. Other vegetables are cooled in tanks of ice water or with sprays of ice water.

Transportation equipment has undergone radical changes. Rail cars refrigerated with water-ice are rapidly being replaced by mechanically refrigerated cars with thermostat controls that can provide the optimum temperature for the particular product. A large volume of produce is being shipped by truck direct from the packing shed to the wholesale distributor. The truck schedules are often as fast or faster than transportation by rail. In recent years there has been an increased use of refrigerated trailer bodies or containers that can be shipped on special rail flat cars and then hauled directly from the rail terminal to the distributor's warehouse. The use of trucks and containers reduces the handling and exposure of the product and speeds its delivery to the consumer. The use of pallets and forklift trucks has speeded the loading and unloading of rail cars and trucks and the handling of agricultural products in packing sheds and warehouses.

Much of the fresh produce is wrapped in plastic films for self-service marketing. Some of the packaging is done at the retail store or distributor's warehouse, but an increasing amount is being done in the packing shed at the shipping point. Packaging reduces losses of moisture and prevents wilting. With some leafy products, such as lettuce, this may aid in reducing loss of vitamins during marketing. Since 1939 there has been a steady decrease in the per capita consumption of fresh fruits and vegetables and an increase in the consumption of processed products. Consumption of frozen foods has increased more rapidly than canned foods, but canned foods still make up about 80 percent of the processed foods.

I do not have figures on how the vitamin contents of fruits and vegetables marketed three decades ago compare with those available in the market today, but they could be expected to be somewhat better now because of the more rapid handling and better refrigeration at all steps. There is considerable data to show that vegetables lose vitamin C more rapidly at room or higher temperatures than when refrigerated. There is also data to show that vitamins decrease with length of time in storage; for instance:

The loss of vitamin C in lettuce was 50 percent in 24 hours at room temperatures, but less than 48 percent in 72 hours when stored in the refrigerator.

In spinach, the loss of vitamin C was 70 percent in 24 hours at room temperature, but only approximately 50 percent after 48 hours in the refrigerator.

Green beans lost 25 percent of their vitamin C in 72 hours at room temperature, but only 5 percent after 96 hours in an ice refrigerator and 10 percent on the shelf in a mechanical refrigerator.

With modern efficient handling, transportation, and storage methods, it is possible to offer the consumer an abundant and varied assortment of fresh, canned, and frozen fruits and vegetables that supply adequate amounts of vitamins if properly used. To reap all of the benefits of this abundance, it is important that the consumer be informed about the proper care and preparation of these foods in the home. Failure to store fresh produce promptly under humid conditions in the refrigerator and keeping too long can result in serious losses of vitamin C. Trimming the dark green outer leaves of lettuce and cabbage results in a serious loss—since they contain 30 times as much vitamin C as the bleached inner leaves.

TOXICANTS NATURALLY PRESENT IN FOODSTUFFS

Another aspect of plant composition which relates to nutritional value is the presence of natural toxins. The content of such substances is generally a heritable characteristic and may be affected by selections made in breeding programs. In some cases, content of toxic material may be enhanced by soil composition or by the growth conditions to which the plant is subjected.

During the centuries of man's existence, he has learned to reject food usage of poisonous species and has selected other, more innocuous, ones for agricultural development and domestication. Yet even these foods are not necessarily perfect for our needs and frequently contain harmful or potentially harmful substances. Some of the natural toxicants may be as damaging as certain of the food additives, pesticide residues, and incidental contaminants to which objections have been voiced.

Cooking or some other processing often eliminates toxic components or reduces their effects. Our best bet for minimizing effects of toxicants that still persist is to eat a varied diet so that we will consume only reasonable amounts of any one food and thus stay below harmful limits of particular toxic constituents that may be present. A number of excellent reviews are available on naturally occurring toxicants in foods. Some examples of naturally occurring toxicants, or potential toxicants, are provided below.

ERUCIC ACID

Erucic acid is a major constituent of rapeseed and mustard seed oils. Though these are not general food ingredients in the United States, they are widely used in various other areas of the world for this purpose—rapeseed is the fifth most important oilseed in world commerce and accounts for about 8 percent of the world's total vegetable oil supplies. Erucic acid is a 22-carbon monounsaturated acid (Table 17). Though no harmful effects on humans have ever been attributed to these high erucic oils, animal experiments have made them suspect and recently rapeseed oil high in erucic acid when fed to experimental animals in substantial amounts, far in excess of normal human consumption, *caused changes in the heart tissue* of some of the animals.

Rapeseed is Canada's most important oilseed crop. As a result of the adverse nutritional effects of the high erucic oils on experimental animals, Canada considers it prudent to change to new rapeseed varieties from which the erucic acid is removed by genetic manipulation. Fortunately, these varieties are already available and have suitable agronomic characteristics. They were developed because high erucic oils do not yield margarines of optimum consistency. We presume that other rapeseed-producing countries will follow Canada's lead. The differences in composition of the new genetic variant from those of present rapeseed oil are shown in Table 17. Craig states that the time required to develop a new variety is only 7 to 8 years instead of 15 to 20 years because of accelerated procedures now used by plant breeders.

GOITROGENIC THIOGLUCOSIDES

Vegetables and condiments of the plant family that includes cabbage, turnips, rutabaga, and mustard contain substantial quantities of compounds called thioglucosides, or glucosinolates (Fig 9). About 50 or 60 related compounds having this structure, with different organic "R" (radical) groups, have now been identified. These substances are acted on by naturally occurring enzymes to give different types of hydrolytic products as shown in Fig 9. Some of these products are strongly goitrogenic. The goitrogenic activity of thiocyanate ion can be counteracted by suitable quantities of iodine; the activity of the oxazolidinethiones cannot be overcome by iodine. Typical thiocyante contents found in some fresh vegetables are shown in Table 18.

Among the leafy and root vegetables consumed by humans, there is variability in thioglucoside content but varieties have not been specifically developed for the purpose of minimizing quantities of goitrogens. Endemic goiter due to foodstuffs is not widespread, but has been noted in areas where the diet is rich in crucifers due to habit or periods of food restrictions during wartime, and especially when the diet as a whole is marginal in iodine content. In some cases, the goitrogenic effect of cabbage-family vegetables is probably reduced by cooking since the enzymes are destroyed that hydrolyze the thioglucosides. However, this point has not been extensively researched. The physiological effects are not necessarily eliminated since man's digestive tract also provides sources of the hydrolytic enzymes from the microbial flora located there.

Seed meals from cabbage-family plants, principally rapeseed, are obtained as a by-product of edible oil production and are used as animal feed. Special processing is required to reduce goitrogenic and other effects of the naturally occuring thioglucosides present. Recently, Scandinavian and Canadian plant breeders have found a strain of rapeseed that is extremely low in thioglucoside content. A vigorous breeding program to incorporate this genetic characteristic into production varieties offers encouragement that goitrogens can be eliminated in the future from cabbage-type vegetables and related oilseeds from which animal feeds are derived. Other types of naturally occurring goitrogens have been reported in soybeans and several other common vegetables, but the principal source of natural goitrogens is in the cabbage-family plants.

47

SOLANINE

It has long been known that under rather well-defined circumstances the white potato could contain toxic amounts of solanine, a glycoalkaloid. Normally present in tubers at levels undetectable even by sensitive tasters, it can exceed 20 mg percent in tubers that are sufficiently exposed to light at temperatures that cause them to accumulate chlorophyll. Thus, the folk warning against greened potatoes has a real basis. Occasionally reports appeared in Europe of poisoning of stock fed waste potatoes and of prisoners and others on diets of high potato content.

Like other members of the Solanaceae, potatoes have relatively high concentrations of the toxic alkaloids in the leaves and flowers, but potatoes have very little in the tubers. No evidence of translocation was found. Other closely related tuber-bearing plants, however, are known to have high levels in their tubers. Twenty-five years ago it was noted that the possibility of new potato varieties accumulating alkaloids should be kept in mind. The reason for this is that breeders often look outside the white potato species for sources of disease or insect resistance.

Surveys of the solanine content are now being made, both of older varieties and breeding materials. Table 19 shows typical values for total glycoalkaloids in some of the most widely grown potatoes. These are average values for potatoes grown in all parts of the country. A level of 20 mg percent is considered questionable. Table 20 shows some single sample values for newly released varieties.

The potato breeder is well aware of the possibility of breeding in "people resistance" when he attempts to increase the resistance of potatoes to insects or diseases by selective breeding experiments. Where such resistance has a chemical or biochemical basis rather than a physical basis, it behooves breeders of all food crops to be aware of the trap in which new varieties may be tailored to be insect resistant and at the same time become toxic to animals or humans.

DIETARY IMPROVEMENT THROUGH PLANT BREEDING FOR INCREASED VITAMIN CONTENT

It has been demonstrated that by plant-breeding programs we can increase the content of ascorbic acid and carotene in several vegetable and fruit crops. Because the food selections of people vary widely, it would be necessary to increase the level of these vitamins in a considerable number of food crops in order to improve the dietary intake of all population groups. However, development of a new commercially acceptable vegetable variety with increased ascorbic acid content, for example, at the maximum present in current breeding lines may require from 8 to 10 years and considerable research investment.

Hence, it is desirable to select those food crops for development which would have maximum impact on dietary improvement. Generally, the requirements to be met are that the food crop chosen should be a food that is widely used and will be accepted and that it is a relatively good food source of the nutrient found to be wanting in the diets of a significant proportion of the population.

For the present discussion, let us consider ascorbic acid as the nutrient and examine the relative contributions of the major food sources of this vitamin to the diets of a selected population group. This will illustrate some of the problems met in selection of a food designed for improvement of the diets of a population group.

According to the USDA nationwide household food consumption survey conducted in the spring of 1965, ascorbic acid was one of the three nutrients for which household diets were most likely to fall short of the Recommended Dietary Allowance (RDA) of the Food and Nutrition Board, National Research Council, National Academy of Sciences. Analysis of findings in a special part of the survey, based on a one-day recall of food used by individuals in households, showed that ascorbic acid levels generally averaged above the RDAs for most of the 22 sex-age groups which included individuals from households at all income levels. However, 16 of the 22 sex-age groups of persons in households with annual incomes under $3,000 averaged below the RDA for ascorbic acid.

For the purpose of making a special study of the eating patterns of the elderly, our analysts examined individual records for persons aged 65 years and over in the North Central Region. An examination of the diets within this age group revealed that only 39 percent of the men and 44 percent of the women had ascorbic acid from one day's food at or above the recommended allowance. On the other hand, 45 percent of both men and women failed to meet even two-thirds of the RDA.

As for the other age groups—potatoes, citrus fruits, and tomatoes were the principal sources of vitamin C for those 65 and over (Fig 10). Potatoes were the most frequently used food; 62 percent of the men ate potatoes on the survey day as compared to 25 percent who ate citrus fruit and 10 percent who ate tomatoes. Moreover, members of this age group are substantial users of potatoes.

Although the survey revealed that consumption of some foods decreased in advanced years, use of potatoes by men and women aged 65 years and over held up quite well (Fig 11). It should be recalled that the data for this particular analysis are based on food eaten in only one day, a 24-hour period. It may have been atypical—more or less than usually eaten. Thus, we should not generalize that these intake levels are representative of each person surveyed. But they may be viewed as a conservative approximation of nutrient intakes from food.

It was significant to find that for those men and women who ate potatoes in a day, but obtained under two-thirds of the recommended allowance for vitamin C, about half of this nutrient came from potatoes (Fig 12). It would seem, therefore, that an appropriate way of increasing their vitamin C intake would be to provide potatoes of higher vitamin C content. We have calculated the improvement that would result in diets of all persons 65 years and over, both users and non-users of potatoes, if new potato varieties were developed which contained 2-½ times average current commercial levels or near the maximum found in present breeding lines (Fig 13). Compared with the survey levels in 1965, in which only 39 percent of

the men met the RDA, now 56 percent would achieve it and only 30 percent would be under two-thirds of the allowance. For women, the results are also favorable, but of somewhat lesser degree, an improvement of 54 percent meeting the RDA.

Thus, development of potato varieties with increased ascorbic acid content would have significant impact in improving the vitamin C intake of elderly people and undoubtedly also that of other age groups. Other sources of vitamin C would be needed, however, to meet full requirements of the several sex-age groups, and effort should be given to improvement in the vitamin C content (and other vitamins as well) of other vegetables and fruits, particularly in the form they reach the consumer, and are prepared for the table.

A considerable part of the vitamin content of fruits and vegetables present at harvest may be lost in subsequent handling, transport, and processing operations. Even with the improved handling and packaging methods introduced in recent years, substantial losses may occur and emphasis should be continued on development of methods that will further reduce losses.

Processed foods constitute an increasing proportion of the US diet. More than half of all fruits and vegetables, and practically all cereal grains, reach the consumer in processed form (Table 21). Most other food products are subjected to some processing operation. Opportunity is afforded in the processing operation for addition of vitamins and minerals to compensate for losses that occur and restore the product to a level comparable to, or greater than, that in the fresh product. Such enrichment of selected food products would be a feasible and economical way to improve the dietary intake of vitamins and minerals for a large part of our population.

Table 16.
Compositional Range (mg/100 gm) in Breeding Lines and in Commercial Varieties.

Vegetables	Constituent	Breeding Lines	Commercial Varieties
Carrots	Carotene	0 - 370	64 - 94
Cabbage	Ascorbic Acid	93	50 - 60
Muskmelon	Ascorbic Acid	3 - 61	30
Tomatoes	Ascorbic Acid	50	15 - 25 (50)*
Tomatoes	Beta Carotene	4	0.4
Sweet Potatoes	Carotene	5 - 22	5 - 12
White Potatoes	Acorbic Acid	30	8 - 15

*Commercial Variety Doublerich.

Table 17.
Genetic Variation in Rapeseed Oil,
$CH_3(CH_2)_7CH = CH-(CH_2)_xCOOH$
Oleic Acid, x = 7; Erucic Acid, x = 11.

Acid	Rapeseed B. napus	Rapeseed (Turnip Rape) B. Campestris	Low Erucic Rapeseed
16 Acids	3.5	4.8	3.6
18:0	1.5	1.3	1.1
18:1	22.5	33.3	54.8
18:2	12.2	20.4	31.1
18:3	5.4	7.6	9.7
20:1	14.2	9.4	0.0
22:1	40.4	23.0	0.0

Table 18.
Thiocyanates in Vegetables.

Vegetable	mg/100 gm Edible, Fresh product
Cabbage	3-6
Kale	3-25
Brussels Sprouts	10
Cauliflower	4-10
Kohlrabi	2-3
Rutabaga, Turnip	9

Table 19.
Total Glycoalkaloid Content of Potatoes.

Variety	Average Glycoalkaloid Content
Kennebec	9.7 mg%
Katahdin	7.9
Russet Burbank	7.9
Irish Cobbler	6.2
Red Pontiac	4.3
Wild Solanium Varieties (20)	198.0

50

Table 20.
Total Glycoalkaloid Content of New Varieties of Potatoes.

Variety	TGA	Variety	TGA
Houma	11.0 mg %	Norchip	6.3 mg %
Oromonte	10.0	Ona	5.4
Cascade	9.3	Alamo	4.9
Chiefton	8.5	Penobscot	4.5
Wauseon	7.5	Norgold Russet	4.6
Norland	6.6	Shurchip	3.3

Table 21.
Per Capita Consumption of Fresh and Processed Fruits and Vegetables.

		1939 Pounds	1969 Pounds
Fruits	Total Processed	29.5	55.5
	Fresh	109.6	23.6
Vegetables	Total Processed	58.0	114.1
	Fresh	116.6	97.6

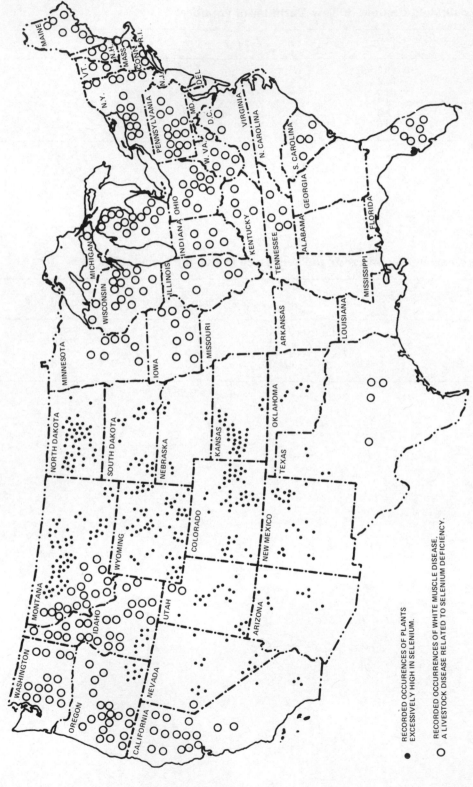

Figure 7.
Areas Where Plant Composition or Nutritional Problems Indicate a Deficiency or Excess of Selenium.

● RECORDED OCCURENCES OF PLANTS
  EXCESSIVELY HIGH IN SELENIUM.

○ RECORDED OCCURRENCES OF WHITE MUSCLE DISEASE,
  A LIVESTOCK DISEASE RELATED TO SELENIUM DEFICIENCY.

Figure 8.
**The Effect of Fertilizer Level, Variety, and Shading
on Ascorbic Acid in Tomatoes.**

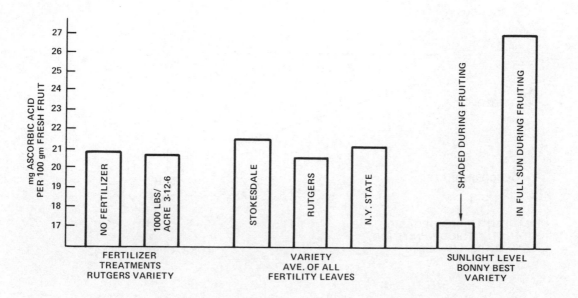

Figure 9.
Typical Glucosinolate Enzymatic Hydrolysis Products.

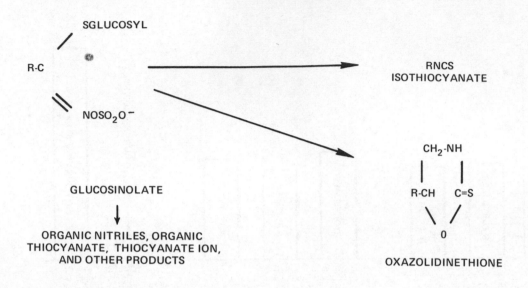

Figure 10.
**Use of Potatoes, Citrus Fruits, and Tomatoes (Persons
65 Years and Older, North Central Region, Spring 1965).**

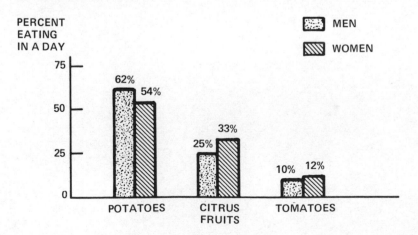

Figure 11.
White Potatoes: Quantity Per Person in a Day.

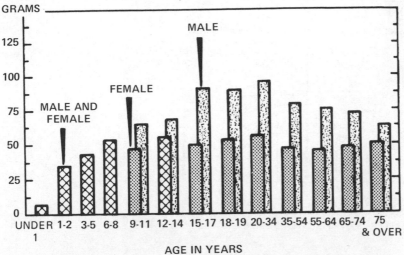

U.S. DIETS OF MEN, WOMEN, AND CHILDREN, 1 DAY IN SPRING 1965.

INCLUDES MIXTURES.

U.S. DEPARTMENT OF AGRICULTURE   NEG. ARS 6011-70 (IQ)

Figure 12.
**Percent of Vitamin C from Potatoes (for Persons Eating
Potatoes in 1 Day, Age 65 and Older, North Central
Region, Spring 1965).***

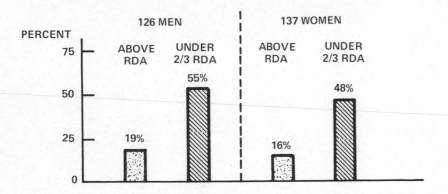

*BY GRADE OF DIET, BASED ON 1968 RDA'S OF NRC-NAS.

Figure 13.
Dietary Effects of Higher Ascorbic Acid Content of
Potatoes (1-Day's Diet, Persons 65 Years and Older,
North Central Region).*

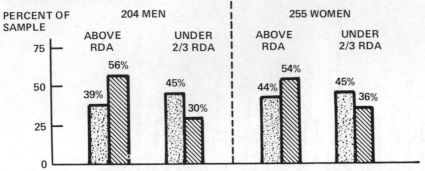

(BEFORE) ▦ SPRING 1965 SURVEY LEVELS.

(AFTER) ▨ ESTIMATED LEVEL IF POTATOES EATEN HAD 2-1/2 x VIT. C.

*BASED ON 1968 RDA'S OF NRC-NAS.

Considerable information is available on the effect of processing on the vitamin and mineral content of foods. Much less information is extant on the biological availability of vitamins and minerals in processed foods. A brief review of the current state of knowledge will be presented, emphasizing the extent of losses or preservation of nutrient properties which processing techniques can influence. Some recommendations will be given at the end of this paper which could act as guidelines to improve the nutritional qualities of processed foods.

The subject matter of this volume is very timely with the growing emphasis on nutrition in the communications media and the increase of awareness of the American public in nutrition. Now it is a generally accepted fact that there is sufficient food in the United States to provide adequate nutrition for the entire population. It is also an accepted fact that there are malnourished people in this country despite the adequacy of the food supply. Since the beginning of civilization, the availability of food has been a continuing concern of man. To prepare food in times of seasonal plenty, to provide a full stomach in the off-season has occupied the time, imagination, and skill of many. Early man preserved his food supply by smoking, curing, and dehydration. These basic methods are still with us today. Added to these are the important commercial processes of canning (thermal processing); dehydration—ie, drum, air, spray, puff and freeze-drying; freezing; fermentation; and refrigerated storage of fresh products.

Fundamental to any processing step is the quality of the raw material to be processed. Attention to quality begins with the selection of the seed, making certain that the variety chosen is not only the most productive in field yield but, also, is the most nutritive from the solids and vitamin content. In the area of vitamin improvement by genetic means, we can mention a few examples. Anderson et al (1954) observed a range of 1.8 to 29.3 mg ascorbic acid per 100 gm in 240 samples of fresh pressed tomato juice. Burrell et al (1940) found a 3-½ fold range of ascorbic acid in 31 strains of cabbage. Bell et al (1942) found a 35 fold range among seven varieties of Muscadine grapes. Marked difference in ascorbic acid content presumably due to genetic variation may be noted by comparing results by Cravioto et

chapter 5

influence of processing on vitamin-mineral content and biological availability in processed foods

by
Richard E. Hein, PhD
and
Imri J. Hutchings, PhD

al (1945) for acerola from Mexico (8.4 mg percent) with results by Asenjo et al (1946) for acerola from Puerto Rico (1707 to 2963 mg percent).

Several years ago, one of the seed companies had a squash selection in which they had little interest, but from the standpoint of nutrition it was of interest to us as a major factor in supplying vitamin A to an infant's diet. Through selective breeding, we were able to increase the vitamin A content tenfold and, in so doing, also increased the vitamin C content so that strained squash became a major contributor of both of these vitamins.

The nutrient content of edible plants is largely under genetic control. In adapting edible plants to mechanical harvesting, let's not lose sight of the basic nutrients contributed by these plants. Let's establish a working relationship between the geneticist (plant breeder) and the food scientist to assure proper nutrient levels in varieties of edible plants under development so that when released, the variety will not only produce an acceptable field yield of good physical quality but will also give an increased nutritive yield.

With a nutritionally acceptable raw product, either the consumption or the processing of that product should occur at the same time that it has its optimum food characteristics. If the product has already undergone degradative changes, these nutritive losses will worsen either on further storage or on processing. Early man recognized that his survival depended on a choice of proper foods, which choice was based on the senses of appearance, odor, texture, and taste of these foods. Nowadays, an attractive food is not necessarily a nutritious food; however, the nutritional quality of a raw product and the food processed from it is related to its general attractiveness in terms of desirable color, lack of decay and bruising, proper trimming, and maturity—as well as on the genetic characteristics.

The effect of processing on the vitamin content of foods is great because vitamins are quite sensitive to a number of variables. In Table 22, the effect of pH, air or oxygen, light and heat on a number of vitamins is listed. The symbol S means stable or no significant destruction, whereas U designates unstable. It will be noted that vitamin C is relatively unstable, especially in the presence of occluded air and heat. Riboflavin is very sensitive to light and the rate of destruction increases as the pH and temperature increase. Because of these sensitivities, cooking losses of some essential nutrients may be in excess of 75 percent. In modern commercial food processing operations, however, losses are seldom in excess of 25 percent.

The losses of mineral salts in processing are solely due to leaching during processing steps. The lack of effect of other variables on the mineral content of foods is indicated in Table 22. The stability of the essential amino acids and fatty acids to variables is also listed here.

In the processing of raw products, high-temperature short-time blanches are most likely to retain the water soluble nutrients and to minimize the loss of heat labile vitamins such as C and B_1. In Table 23, data on the retention of vitamins B_1, B_2,

C, and niacin on five vegetables blanched for commercial processing are listed. Blanching is used for the destruction of enzyme systems such as the enzyme ascorbic acid oxidase. Steam blanching of spinach for 2 to 2¾ minutes permitted retention of the four vitamins listed ranging from 100 to 88 percent. Water blanching in Draper-type equipment for 2¾ to 5 minutes at temperatures in the range of 160°F to 20°F resulted in a retention of these four vitamins ranging from 64 to 95 percent with ascorbic acid being affected to the greatest extent. Immersion blanch of spinach for 45 minutes in 160°F water gave the 6 percent vitamin C retention figure (oxidation and leaching). Other factors being equal, steam blanching extracts about half as much vitamin C as hotwater blanching.

Table 24 shows the retention of vitamin C in tomato juice that is heated in the canning process. In tomato juice and in many other canned products, the exclusion of oxygen and trace metals, particularly copper, is important to prevent extensive vitamin C loss. While heat processing of foods does have a negative effect on vitamins, it should also be noted that heat processing has the important merits of destroying deleterious factors in foods such as the antitryptic factor in soy bean and avidin in eggs. It thereby improves their nutritional quality. While heat processing can lower the biological value of proteins by reducing the availability of sulfur-containing amino acids and lysine, a beneficial side effect of this combination of amino acids with reducing sugar is the increased palatability and visual attractiveness of the product.

In Table 25, the vitamin losses in cooking water for a number of vegetables are given. It is, of course, important to avoid this loss of nutrients in the liquid phase to as great an extent as possible in processing. Kramer and others have reported that about one-third of the total water soluble nutrients are found in the liquid phase. The high-temperature short-time processes (HTST) and particularly the agitated cooker-type which increases the rate of heat transfer into the cooked product, are preferred for nutrient preservation as well as for other product acceptability characteristics such as flavor and color. Goldblith, in a presentation at the National Canners Association in 1971, pointed out the immediate foregoing statement is based on the fact that a 10°C rise in temperature results in a tenfold bacterial destruction while approximately doubling the loss of thiamine.

The retention of vitamins in commercially packed juice during storage is described in Table 26. The effect of the higher storage temperature on the vitamin C content may be noted by comparing values for the three temperatures. Also, storage of canned orange juice for 12 and 14 months at the same temperature shows the additional loss of vitamin C for the longer storage period.

The blanching of vegetables—prior to dehydration processes—is important if losses of vitamin C and vitamin A through enzymatic activity are to be avoided. Only moderate losses of thiamine, riboflavin, niacin, and pantothenic acid occur during dehydration. Results of several investigations are summarized in Table 27. Dehydration, in most cases, reported in this table was done in the forced air, cross-flow tunnel driers. Vacuum puff drying and freeze-drying generally preserve the vitamin content of the product to a large degree. In fact, fresh varieties of foods are likely to lose more nutrients by wilting and bruising than by freeze-drying or by freezing soon after harvest.

The freezing process itself does not impair the vitamin content of the product. Again, it is important that blanched vegetables be frozen with a minimum of delay to avoid microbiological changes which could affect nutritive values, flavor, and appearance. The storage of frozen foods at 0°F or at lower temperatures results in negligible vitamin loss over several months. However, with storage at 16°F, losses of vitamin C in frozen asparagus, peas, and lima beans exceeded 50 percent after six months of storage. A comparison of processed (canning, air-drying, freeze-drying, and freezing) on vitamin C content of peas is given in Table 28. This information is taken from Bender (1966). Please note the loss of ascorbic acid on cooking. One of the most important factors in the destruction of preservation of the vitamin and nutrient content of foods is the final preparation of foods in the home and the restaurant, school cafeteria, etc. The steam table can destroy much of what had been preserved in all prior steps.

The brining and fermentation of vegetables is an old preservation technique still used in modern processing for specialty products. While cucumber pickles are considered mainly as condiments, they do possess some nutritive value. Dill pickles retain 33 to 60 percent of the ascorbic acid originally present. Practically all the carotene is retained in the fermented dill pickle.

The changes which may occur in refrigerated storage of foods are many. Factors such as growing conditions and varieties of plants, feeding practices for animals, conditions of harvest and slaughter, sanitation and damage to tissue and, of course, temperature and conditions of storage can all influence storage changes. As example of these factors, pigs fed substances high in unsaturated fats, such as soybeans and peanuts, produce softer pork and lard than the same pigs fed on corn and other cereal grains. Flesh of the corn-fed pigs keeps better in cold storage. Animals that are exercised or excited before slaughter use part of their glycogen reserves—thus, less glycogen is available for conversion to lactic acid and storage qualities are impaired.

The effect of storage temperature and time on the vitamin C content of asparagus, broccoli, green beans, and spinach is given in Table 29. Other data show the loss of thiamine and riboflavin for storage of asparagus for one week at 32-40°F was 18 percent and 14 percent, respectively. It should be noted that at least in recent years, the term "fresh" as applied to vegetables, fruits, etc., signifies that the product is unprocessed rather than that the product has been recently harvested. As a final remark on refrigerated storage, while refrigerated storage temperatures may sometimes be lower than optimum, for the majority of perishable foods these temperatures are still better than no cooling at all.

The effect of processing on vitamin levels in foods must be viewed in terms of the importance of that food's contribution to the minimum daily requirement for that particular vitamin. For example, the loss of ascorbic acid in milk on exposure to sunlight is not really consequential because the contribution of milk to the RDA of vitamin C is not very important. It may be meaningful, therefore, to look at both the per capita intake of processed and fresh products as well as the nutrient contribution of these food categories.

In these last three tables the per capita consumption of food categories for

1968 (preliminary data from USDA's *Agricultural Statistics 1969*) are listed as well as the nutrient contribution of these foods. In Table 30, the pounds of food expressed as the retail weight equivalent expressed as the average consumption per person in the United States is listed. Butter is included in with the animal fat category (5.6 lbs). Pounds of fresh fruit consumed per capita have decreased about 20 percent in the last 14 years with about a 10 percent increase in processed fruit. Similarly, fresh vegetable consumption has decreased with an increase in processed and frozen products.

Please note in Table 31 the percent contribution of the 10 food groups to the total calories, protein, fat, carbohydrate, and minerals available for consumption per capita per day. The meats and the flour and cereal product categories make substantial contributions as do the dairy products. In Table 32, the percent contribution of this same 10-food group to the vitamin picture is given. Citrus fruits contribute about two-thirds of the ascorbic acid in the fruit category—tomatoes contribute substantially to vitamin C in the vegetable category. The fortification of food products with vitamins and minerals to compensate for those nutrients which may be lost in processing or storage before consumption is worthy of consideration. Certainly, juices which replace orange juice should be fortified with vitamin C to the average level of vitamin C found in orange juice at the time of consumption. The question of fortification of certain products with B vitamins either because of processing losses or because of diet substitution also needs considerable thought and discussion.

The relationships of vitamins with other nutrients and within themselves is somewhat complex. For example, all members of the "B-complex vitamins" are needed for optimal glucose absorption in the rat. Disturbed carbohydrate metabolism was noted in pantothenic acid deficiency. Protein utilization seems to be affected by riboflavin and pyridoxine deficiencies. It has been observed that an excess of protein aggravated the symptoms of a vitamin B_{12} deficiency. The biological availability of iron probably is dependent on its chemical form to some extent, although further work is needed to define the relation of the compound containing the iron and its utilization. The biological availability of most of the minerals does not appear to be dependent upon their chemical state in the food we eat.

The foregoing remarks provide a brief review of the present state of knowledge of the influence of processing on vitamins and minerals and their availability in processed foods. The geneticist and grower must be encouraged to emphasize the nutritional properties of the plant and amimal rather than yield and appearance consideration only. Great effort is needed to preserve the nutrient qualities by correct raw material handling, minimizing blanching losses, rapid heat processing, good packaging, and proper storage conditions. Research efforts in several fields are needed to optimize the nutrient contents and balance of processed products. Finally, it is evident that proper communications of the current nutritional values of processed foods to the consumer is of paramount importance.

Table 22.
Stability of Nutrients. [*]

Nutrient	pH 7	Acid	Alka-line	Air	Light	Heat	Cooking Losses, Range
Vitamins							%
Vitamin A	S	U	S	U	U	U	0- 40
Ascorbic Acid (C)	U	S	U	U	U	U	0-100
·Carotenes (pro-A)	S	U	S	U	U	U	0- 30
Cobalamin (B_{12})	S	S	S	U	U	S	0- 10
Vitamin D	S	U	U	U	U	U	0- 40
Niacin	S	S	S	S	S	S	0- 75
Vitamin B_6	S	S	S	S	U	U	0- 40
Riboflavin (B_{12})	S	S	U	S	U	U	0- 75
Thiamine (B_1)	U	S	U	U	S	U	0- 80
Tocopherols (E)	S	S	S	U	U	U	0- 55
Essential Amino Acids							
Isoleucine	S	S	S	S	S	S	0- 10
Leucine	S	S	S	S	S	S	0- 10
Lysine	S	S	S	S	S	U	0- 40
Methionine	S	S	S	S	S	S	0- 10
Phenylalanine	S	S	S	S	S	S	0- 5
Threonine	S	U	U	S	S	U	0- 20
Tryptophan	S	U	S	S	S	S	0- 15
Valine	S	S	S	S	S	S	0- 10
Essential Fatty Acids	S	S	U	U	U	S	0- 10
Mineral Salts	S	S	S	S	S	S	0- 3

[*]Source—Harris and Loesecke, 1960.

Table 23.
Retention of Vitamins in Vegetables Blanched for Commercial Canning. [*]

Vitamins	Product	No. of Tests	% Retention		
			Max.	Min.	Mean
Ascorbic Acid	Asparagus	26	100	74	95
	Green Beans	38	90	50	74
	Lima Beans	12	83	54	72
	Peas	60	90	60	76
	Spinach	41	99	6	67
Niacin	Asparagus	8	100	77	94
	Green Beans	29	100	60	95
	Lima Beans	8	98	68	81
	Peas	39	96	59	73
	Spinach	34	100	63	83
Riboflavin	Asparagus	12	100	72	90
	Green Beans	29	100	70	95
	Lima Beans	8	100	59	76
	Peas	37	87	67	75
	Spinach	37	100	78	88
Thiamine	Asparagus	12	100	79	92
	Green Beans	34	100	82	91
	Lima Beans	12	77	36	58
	Peas	60	100	63	88
	Spinach	35	100	67	85

[*]Source—Harris and Loesecke, 1960.

Table 24.
Retention of Vitamins in Tomato Juice Heated in Canning Process. *

Vitamin	No. of Tests	% Retention		
		Max.	Min.	Aver.
Ascorbic Acid	90	90	35	67
Thiamine	18	100	73	89
Riboflavin	17	100	86	97
Niacin	17	100	83	98
Carotene	7	74	60	67

*Source—Harris and Loesecke, 1960.

Table 25.
Vitamin Losses into Cooking Water. *

Vegetable	Cooking Time	Thiamine	Riboflavin	Niacin	Ascorbic Acid
	min.	%	%	%	%
Corn (canned)	30	22	27		
Carrots (canned)	30	19			12
Potato	25				5
Asparagus (canned)	30	22	22		23
Bean (canned)	30	31	30		23
Broccoli	5-13	17-22	11-14		10-12
Broccoli	9-16	40-46	41-54		16-35
Cabbage	20-120	46-72	53-78		26-33
Cabbage	4-8	24-35	20-53		14-38
Lima Beans (canned)	30	22	19		16
Peas	9				9-14
Spinach (canned)	30	26-32	23		21-28
Corn		15	12	14	30
Carrots (dehy.)	25-45	52-93	55-89	42-58	

*Source—Harris and Loesecke, 1960.

Table 26
Retention of Nutritive Factors in Commercially Packed Juices During Storage. *

Juice	Storage Temperature °F	Storage Period Months	Ascorbic Acid	% Retention		
				Niacin	Riboflavin	Thiamine
Grapefruit	50	12	95	—	—	99
	65	12	91	—	—	100
	80	12	75	—	—	93
Orange	50	12	97	—	—	100
	65	12	92	—	—	98
	80	12	77	—	—	89
Orange	50	12	95	—	—	101
	65	24	80	—	—	94
	80	24	50	—	—	83
Pineapple	50	12	110	—	—	93
	65	12	108	—	—	93
	80	12	93	—	—	87
Tomato	50	24	102	94	92	103
	65	24	92	97	94	94
	80	24	74	98	94	77

*Source—Harris and Loesecke, 1960.

Table 27.
Average Retention of Thiamin, Riboflavin, Niacin, and Pantothenic Acid in Vegetables during Dehydration, Expressed as Percentage of Content in the Fresh Vegetable Before Blanching. *

Vegetable	Thiamine Blanched	Thiamine Dehydrated	Riboflavin Blanched	Riboflavin Dehydrated	Niacin Blanched	Niacin Dehydrated	Pantothenic Acid Blanched	Pantothenic Acid Dehydrated
	%	%	%	%	%	%	%	%
Snap Beans	93	88	93	89				
Beets	84	80						
Carrots	83	59	98	92	93	93	100	94
Sweet Corn	95	91	96	91				
Onions	—	84	—	95	—	88	—	91
Green Peas	93	90	92	87				
Potatoes	80	60	100	100		94	100	100
Potatoes	68	77						
Sweet Pot.	—	—	—	89	—	100	—	72
Rutabagas	86	83						

*Source—Harris and Loesecke, 1960.

Table 28.
Retention of Ascorbic Acid in Peas at Various Stages of Processing and After Cooking. [*]

After Step	Process				
	Fresh	Frozen	Canned	Air-Dried	Freeze-Dried
	Retention of Vitamin C (Per Cent)				
Blanching	—	75	70	75	75
Processing	—	75	63	45	40
Thawing	—	71	—	—	—
Cooking	44	39	36	25	35

[*]Source—Harris and Loesecke, 1960.

Table 29.
Losses in Vitamin C in Selected Vegetables on Cold Storage. [*]

Produce	Storage Days	Conditions Temp. °F	Losses, %
Asparagus	1	35	5
	7	32	50
Broccoli	1	46	20
	4	46	35
Green beans	1	46	10
	4	46	20
Spinach	2	32	5
	3	34	5

[*]Source—Harris and Loesecke, 1960.

Table 30.
Per Capita Consumption—1968.

	Pounds		Pounds
Meat	161.9	Fresh Fruits	76.7
Poultry	45.4	Processed Fruits	51.8
Fish	14.0	Fresh Vegetables	139.6
Eggs	40.5	Canned Vegetables	50.7
Dairy Products	357.0	Frozen Vegetables	9.1
Fats and Oils		Potatoes (inc. Sweet)	105.7
Animal	17.9	Dry Beans and Peas	16.3
Vegetable	34.8	Flour (cereal Prod.)	143.0
		Sugar and Sweeteners	114.8

Table 31.
Macronutrients and Minerals: Percentage of Total Contribution by Food Groups—1968.

Food Group	Food Energy	Protein	Fat	CHO	Ca	P	Fe
Meat, Poultry, Fish	20.3	40.5	35.2	0.1	3.5	25.3	30.1
Eggs	2.3	5.9	3.4	0.1	2.6	6.0	6.1
Dairy Products	11.8	22.6	13.4	7.2	76.2	36.7	2.2
Fats and Oils	17.0	0.1	40.7	Tr	0.4	0.2	0
Fruits	3.2	1.1	0.3	6.7	2.0	1.8	4.6
Potatoes (inc. Sweet)	2.9	2.5	0.1	5.6	1.0	4.1	4.5
Vegetables	2.7	3.7	0.4	5.1	6.2	5.5	11.2
Dry Beans and Peas	2.9	5.0	3.6	2.1	2.6	5.7	6.4
Flour (Cereal Prod.)	19.9	18.2	1.4	35.9	3.4	12.5	26.8
Sugars and Sweeteners	16.3	Tr	0.0	36.4	1.1	.2	5.5
Total	100%	100%	100%	100%	100%	100%	100%
•	3200	98g	150g	371g	0.93g	1.51g	16.9mg.

•Quantities available for consumption per capita per day—1968.

Table 32.
Vitamins: Percentage of Total Contribution by Food Groups—1968.

Food Group	Vitamin A	Thiamine	Riboflavin	Niacin	Ascorbic Acid
Meat, Poultry, Fish	22.9	29.4	24.6	46.0	1.1
Eggs	6.8	2.5	5.9	0.1	0
Dairy Products	11.8	9.9	43.1	1.7	4.7
Fats and Oils	8.6	0.0	0.0	0.0	0.0
Fruits	7.3	4.3	2.0	2.5	35.0
Potatoes (inc. Sweet)	5.7	6.7	1.9	7.6	20.9
Vegetables	36.4	8.0	5.6	6.8	38.3
Dry Beans and Peas	TR	5.5	1.8	7.0	Tr
Flour (Cereal Prod.)	0.4	33.6	14.2	22.7	0.0
Sugar and Sweeteners	0.0	Tr	0.1	Tr	0.0
Total	100%	100%	100%	100%	100%
•	7700 IU	1.82 mg	2.24 mg	22.1 mg	107 mg

•Quantities available for consumption per capita per day—1968.

Agricultural Statistics. US Department of Agriculture, 1969.

Anderson, E. E., Fagerson, I. S., Hayes, R. M., and Fellers, C. R. "Ascorbic acid and sodium chloride content of commercially canned tomato juice." *J. Am. Diet. Assoc.* 30: 1250, 1954.

Asenjok C. F. and de Guzman, F. "The high ascorbic acid content of the West Indian Cherry." *Science* 103: 219, 1946.

Bell, T. A., Yarbrough, M., Clegg, R. E., and Sotterfield, G. H. "Ascorbic acid content of seven varieties of Muscadine grapes." *Food Res.* 7: 144, 1942.

Bender, A. E. "Nutritional effects of food processing." *Jour. Food Tech.* 1: 261, 1966.

Brush, M. K., Hinman, W. F., and Halliday, E. G. "The nutritive value of canned foods. V. Distribution of water soluble vitamins between solid and liquid portions of canned vegetables and fruits." *Jour. Nutr.* 28: 131-140, 1944.

Burrell, R. C., Brown, H. D., and Elright, V. R. "Ascorbic acid content of cabbage as influenced by variety, season, and soil fertility." Food Res. 5: 247, 1940.

Cameron, E. J., Clifcorn, L. E., Esty, J. R., Feaster, J. F., Lamb, F. C., Monroe, K. H., and Royce, R. *Retention of Nutrients during Canning.* Research Laboratories, National Canners Association, 93 pp, Washington, D.C., 1955.

Cameron, E. J. and Esty, J. R., (eds). *Canned Foods in Human Nutrition.* National Canners Association, 264 pp, Washington D.C., 1950.

Cecil, S. R. and Woodroff, J. G. "The stability of canned foods in long-term storage." *Food Tech.* 17(5): 131-138, 1963.

Cravioto, R., Lockhart, E. E., Anderson, R. K., Miranda, F. de P., and Harris, R. S. "Composition of typical Mexican foods." *J. Nutrition* 29: 317, 1945.

Feaster, J. F., Tomkins, M. D., and Ives, M. "Influence of processing technique on vitamin retention." *Information Letter,* National Canners Association 1200: 108-109, 1947.

Goldblith, S. A. "Factors which may favorably or adversely affect nutrition in the processing of food." Presentation of National Canners Assoc. Meeting 1-27-71.

Harris, R. S. and Von Loesecke, H. *Nutritional Evaluation of Food Processing.* John Wiley & Sons, Inc., 612 pp, New York, N.Y., 1960.

Jansen, G. R., Ehle, S. R., and Hause, N. L. "Studies on the nutritive loss of supplemental lysine in baking. I. Loss in a standard white bread containing 4% nonfat dry milk." *Food Tech.* 18(3) : 109-113, 1964a.

Jansen, G. R., Ehle, S. R., and Hause, N. L. "Studies on the nutritive loss of supplemental lysine in baking. II. Loss in water bread and in breads supplemented with moderate amounts of nonfat dry milk." *Food Tech.* 18(3): 114-117, 1964b.

Kohman, E. F. *Vitamins in Canned Foods, Bulletin 19L,* Research Laboratories, National Canners Association, Washington, D. C., 1922.

Kramer, A. "The Nutritive value of canned foods. VIII. Distribution of proximate and mineral nutrients in the drain and liquid portions of canned vegetables." *Jour. Am. Diet. Assoc.* 21: 354-356, 1945.

Lincoln, R. E., Zscheile, F. P., Porter, J. W., Kohler, G. W., and Caldwell, R. M. "Provitamin A and Vitamin C in the genus Lycopersicon." *Botan Gaz.* 105: 113, 1943.

Livingston, G. E., Esselen, W. B., Feliciotti, E., Westcott, D. E., and Baldauf, M. P. "Quality retention in baby foods processed by high temperature-short time methods." *Food Tech.* 11(1): 1-5, 1957.

Teply, L. J., and Derse, P. H. "Nutrients in cooked frozen vegetables." *Jour. Am. Diet. Assoc.* 34: 836-840, 1958.

INTRODUCTION

The storage life of processed and pack-aged foods is determined by the reactivity of the food with substances or influences which are capable of causing deterioration, and the ability of the package to protect the food from such influences. Modern preservation techniques have extended the storage life of foods from a matter of days to a period of months or years. Despite this remarkable extension in food storage life, consumer acceptability is the prime factor which limits storage life and unacceptable quality may develop at a time when the nutrient value is still significant. There is no practical value in extending nutrient retention storage studies beyond the time during which the food may be expected to have quality acceptable for consumption.

An exception to this exists in the case of foods intended for long-term use under emergency conditions which may be associated with military or civilian needs in the event of either natural- or war-created catastrophes. Such long-term storage studies, ranging up to seven years, have been conducted by the US Army Laboratories and the University of Georgia.[1-4] They found an apparent "aging" effect upon quality which varied appreciably for different products. Thiamine and vitamin C content decreased faster during storage than did either color or palatability. Conversely, none of the relatively minor carotene losses was as great as the preceding or accompanying decreases in quality ratings. The results showed that there was accumulative damage to both tin cans and flexible containers caused by excessive moisture and heat, and it was concluded that selection or development of suitable containers can be more important than special formulation of foods for long-term storage.

No general statement can be made on the relation of nutritive value to quality. Some nutritive factors are not associated with quality, while those that are so associated show correlations so far from perfect as to make it obvious that factors other than those which influence nutritive value make more significant contributions to quality. Nutritive factors having close relationships to quality may show either positive or negative correlations.[5] Foods are complicated mixtures of chemicals. The known nutrients are only

chapter 6

influence of storage and distribution upon vitamin-mineral content and biological availability in processed foods

by
Ira I. Somers, PhD
Richard P. Farrow
and
James M. Reed

some of these. By far the greatest number of the food chemicals may be of no metabolic use to man, providing only color, flavor, odor, form, or consistency to our food.

Over 70 percent of the food eaten by man today has been subjected to some form of processing, and such processed products include not only individual foods but also pre-packaged meals and entrees. Data compiled by the Economic Research Service of USDA[6] reveal that in 1967 canned and frozen fruits and juices represented 3.3 percent of total per capita food consumption; while canned and frozen vegetables represented 3.5 percent; baby food and soup represented 1.2 percent; and cereal products represented 6.9; with the remaining 85.1 percent consisting of animal products, vegetable oils, sugar, and fresh and dried fruits and vegetables.

Once the food has been treated the subsequent vitamin losses are small if the food is kept cool and, in the case of dry foods, if the moisture content is kept low. The exclusion of air during storage is beneficial since this inhibits oxidative rancidity and protects vitamins A, D, and E —as well as vitamin C which is destroyed by oxidation.

This paper will center on nutrient retention during storage of canned and frozen foods, which by the preceding figures totaled 8 percent of the per capita food consumption in 1967. However, per capita consumption of individual fruits, fruit juices, and vegetables in the canned and frozen form may range from about 20 percent to 90 percent, with the remainder being consumed in the fresh state. The general concepts discussed are applicable to the storage of all processed foods.

CANNED FOODS

Storage temperature An intensive study of food composition and retention of nutritive quality during processing and storage of canned foods was begun in 1942 under the joint sponsorship of the National Canners Association and the Can Manufacturers Institute. Two general types of retention studies were made; vitamin retentions during the complete canning procedure, and retentions during particular canning operations, including storage and distribution. These studies generated 42 published papers covering separate investigations and were discussed in two reviews.[7] Comprehensive compiliations of the data were published by the National Canners Association.[8,9] Eight of the papers dealt with the effects of storage.[10-17]

To obtain information on warehouse temperatures, detailed records were obtained at four warehouses in widely separated parts of the country over a two-year period.[16] Two were in "high temperature" areas (New Orleans and Tampa), while one (in Illinois) represented a cooler area where warehouses are heated in the winter, and one (in California) represented the West Coast valley conditions. In addition, maximum and minimum daily temperature readings were taken in 75 warehouses located throughout the continental United States and in Hawaii.

In most instances average storage temperatures fell within the range of 65°-75°F. The highest temperatures were obtained in Hawaii where the range of the two-year averages was 78°-80°F. In a few of the southern states and in Hawaii average monthly temperatures in excess of 80°F were recorded during the summer. The highest, constant storage temperature (80°F) chosen for the University nutrient retention studies was slightly above the highest average yearly commercial warehouse temperature.

Samples of canned tomatoes, orange juice, and peas, companion to those stored at constant temperatures of 50°, 60°, and 80°F in the University studies, were held for 24 months at nine commercial warehouses. These were analyzed periodically to compare the effect of constant and fluctuating (but similar average) temperatures on the retention of ascorbic acid, thiamine, and carotene. Results indicated that vitamin retentions are essentially the same under constant and fluctuating storage temperatures within the ranges observed commercially.[13-15]

Vitamin and mineral retention Retentions of ascorbic acid, carotene, niacin, riboflavin, and thiamine during storage for up to 24 months are summarized in Tables 29 and 30 of "Retention of Nutrients during Canning."[9] The quantitative results obtained vary for different fruits and vegetables as one would expect for these complicated food and chemical mixtures of differing acidity. However, the qualitative relationships for the retention of some vitamins are highly dependent on storage temperature and time. The commonly accepted rule that chemical reaction rates double with each 18°F rise in temperature can serve only as a rough guide in predicting either nutrient or quality retention. Such predictions appear to be approximately valid for a rather restricted range of food storage temperatures. Extrapolation by this rule to food storage temperatures above 85° to 90°F is not supported by experimental findings.

The percentage retention of ascorbic acid in canned tomato juice stored for 24 months at 50°, 70°, and 85°F and for 12 months at 100°F is shown graphically in Fig 14, taken from Lamb, et al.[17] The juice was commercially packed at one cannery in plain No. 2 cans. Retention was high at 50°F, approximately 98 percent. At 70°F which closely approximates normal storage and distribution temperature, retention at 12 months was 93 percent, and at 24 months was 82 percent. Loss of ascorbic acid was substantially more rapid at 85°F and extremely rapid at 100°F. In this instance, the original juice contained 20.4 mg per 100 gm. Fortification to 70.5 mg per 100 gm resulted in the same percentage retentions at 70°F storage. The retention in fully enameled cans was slightly, but insignificantly, lower than in plain cans.

Lamb[12] also studied ascorbic acid retention in canned grapefruit and orange juice. During storage at 70°F for 18 months, grapefruit juice lost ascorbic acid at a constant rate of about 1 percent a month. Orange juice lost ascorbic acid more rapidly during the first few months of storage and less rapidly during the following months, but the average monthly rate of loss at the end of 12 months was about 1 percent.

Carotene percentage retention figures are given in Table 33 for canned tomato juice and peaches.[9,18] The carotene retentions in tomato juice at all three storage temperatures were good (94 to 98 percent). While the 63 to 64 percent carotene retention for peaches after two years at 65° or 80°F may be considered to represent a significant loss, storage times of up to two years at 80°F had a comparatively small effect upon carotene retention in most products. This is illustrated by the carotene retentions shown in Table 34 for canned carrots and yellow corn.[9,18] With respect to most products, the adverse effect of storage temperature on carotene retention was less marked than was the duration of storage. It should be mentioned that Panalaks and Murray[19] have reported on studies relating to the correlations between crude carotene contents and calculated biopotencies.

Table 35 tabulates riboflavin retention in canned asparagus and sweet peas.[9,18] As in the case of carotene, both riboflavin and niacin retentions appear to be more adversely affected by storage time than by storage temperature.

USDA Handbook No. 8 shows that some canned fruits, vegetables, meats, and specialties may contribute from 6 to 22 percent of the 1.4 mg thiamine RDA in a 100 gm serving.[20] For example, the thiamine levels of canned sweet peas, pineapple, meat and vegetable baby food dinners, pork and beans, and split pea soup are listed as equal to or slightly exceeding the thiamine content of fresh, standard grade carcass beef (0.08 mg per 100 gm). Canned pork luncheon meat is listed as containing 0.31 mg per 100 gm (22 percent of the RDA).

Thiamine analyses after 24 months storage at 80°F shows 72 percent retention in sweet peas and 89 percent retention in pineapple.[14,15] According to Rice,[21] thiamine content of pork luncheon meat after heat processing in 12 oz cans was 0.52 mg per 100 gm. His tests at 80°F storage indicate that the Handbook No. 8 listing of 0.31 mg per 100 gm would represent 8 months storage (60 percent retention), while he found 52 percent retention (0.27 mg per 100 gm) after 10 months storage.

Bongolan, et al,[22] subjected split peas with ham (a commercially prepared baby food) to high-temperature, short-time (HTST) and conventional processes. One lot was enriched with 2.5 mg of thiamine hydrochloride per gm. They reported better percent thiamine retention during 6 months storage at 85°F for the HTST processed product than for that which was conventionally processed. They also found a significant increase in percent retention during storage for the thiamine-enriched product with both processing methods. For example, the thiamine retention after 6 months storage at 85°F was 49.1 percent for the enriched split peas with ham and 25.0 percent for the unenriched product.

The major inorganic elements in foods are the base-forming calcium, magnesium, sodium, and potassium; and the acid-forming phosphorus, chlorine, and sulfur. None of the minerals are altered significantly during the storage of foods canned in either metal or glass containers. However, the total iron content of foods canned in metal containers increases during storage, especially for foods packed in plain cans.

Theriault and Fellers[23] stored fruits and vegetables canned in glass and in plain and "C" enameled, or fruit enameled, cans for up to 200 days at room temperature. There was no change in the total and available iron in the glass pack. The total iron of the foods in cans after storage varied with the pH of the food and the presence or absence of enamel. Rat bioassay was used to determine available iron and iron gained from the cans seemed to be nearly 100 percent available. The percent of total iron shown to be biologically available remained approximately constant throughout the storage period, with availability ranging from a low of 20 percent of the total for canned peaches to 100 percent for canned corn. The biological availability of iron in other specific products was about 50 percent of the total for asparagus, green beans, and spinach, and about 63 percent for lima beans.

Measures to improve vitamin retention Losses of ascorbic acid and some of the other vitamins increase with high storage temperatures and with lengthened storage times. It appears that maintenance of the lowest practicable storage temperature and rapid distribution for consumption would be the most efficacious in achieving maximum retention of nutrients. In 1945 and 1946, Monroe, et al[16] found average warehouse temperatures fell within the range of 65° to 75°F, although 80°F was approached in some instances. Maintenance of a storage temperature not exceeding 70°F would be beneficial.

Reister, et al,[24] reported on steps taken to reduce temperatures in storage warehouses. These included insulation, the use of heat-reflecting paints or materials, and vents or mechanical venting. It is rather common practice to water cool heat-processed canned food to an average temperature of about 100°F so that the residual heat will dry the container to prevent external corrosion. The location within the pallet load affects the time required for the temperature of the can contents to reach equilibrium with the air temperature. At the center of a pallet load this might require as much as three to four weeks, depending upon product, cases, and stacking arrangement. Reister, et al suggested the cooling of processed cans to 70°F, followed by mechanical drying, as one means of avoiding high-storage temperatures during this initial period. The National Canners Association has recommended that water-cooled containers be dried before further handling to prevent spoilage due to micro-leakage. We believe that this is an increasing practice and will permit cooling to 70°F.

In locations where night temperatures are substantially below those in daytime, forced ventilation during the night hours can effectively reduce average warehouse temperature. A precaution to be observed is the avoidance of temperature-humidity conditions which might cause condensation on the cans, jars, or cases. This occurs when the temperature of the containers is at or below the dew point. An excellent discussion of this factor is given by Murray.[25]

Recent statistics show that approximately 70 percent or more of a given season's pack is shipped from the canner's warehouse by the beginning of the new canning season. It is also our understanding that merchandizing and distribution practices have changed to the point where the canner now does most of the warehousing, shipping directly to chain stores or central distribution locations where the canned foods are held for very short periods. Based on this information, it appears that most canned foods reach the consumer well within 18 to 24 months after packing, but it seems doubtful that this storage time can be reduced for many products.

The hermetically sealed containers presently used for canning serve to protect the products against excessive nutritive loss during storage. New containers which may come into use, such as laminated flexible pouches, should be comparable in this respect. The National Canners Association, in its educational program on canned foods, distributes three publications which contain answers to the question, "Where should canned foods be stored?" and "How long will canned foods keep?"

As to storage location, the publications state that the best storage is in a dry place at a moderately cool temperature. It is suggested further that placing the food near steam pipes, radiators, furnaces, or kitchen ranges should be avoided. As to storage time, the publications state that extremely long periods of storage at high temperatures may result in some loss of color, flavor, and nutrient value, and suggest use and replacement at regular intervals. Almost one million of these three publications were distributed in 1970.

FROZEN FOODS

Storage and distribution Time and temperature are two equally important factors in the retention of high quality in frozen foods during storage and distribution to the consumer. The frozen food industry uses the premise that most frozen products will retain good quality for at least one year if held at 0°F or below. This premise is simply supported by the comprehensive series of studies on the time-temperature tolerance of frozen food begun in 1948 by the USDA Western Regional Research Laboratory at Albany, Calif. The general design of the USDA experiments was described in the first of 24 technical papers published in *Food Technology*.[26] These studies provided an enormous amount of quantitative information on the behavior of commercial packs of fruits, vegetables, and poultry products and a few "pre-cooked" items. Results of the USDA time-temperature tolerance studies have been summarized in the 1969 book titled, *Quality and Stability in Frozen Foods*.[27]

Time-temperature surveys were conducted in 1966 by sending time-temperature recording devices through the normal distribution channel and through separate segments of the channel. Data obtained are reported and statistically treated by Byrne and Dykstra in Chapter 13, "Surveys of Industry Operating Conditions and Frozen Product Histories."[27] The authors report the mean time for the total distribution chain to be 5.5 months at an "equivalent temperature" of 3°F. The mean times and "equivalent temperatures" reported are: primary warehouse, 1.95 months at minus 10°F; transportation and distribution to store, 1.3 months at 3°F; and in store and home 2.25 months at 8°F.

They point out that the mean temperature of 3°F for the total distribution chain is this low only because of good primary warehouse temperatures in the minus 10°F range and that there is a "top half" of higher temperatures and times. For example, a survey of 1,610 retail store cabinets for top-level temperatures showed a mean (50 percent probability) temperature of 1°F, while 20 percent were above 9.5°F, and 3 percent were above 20°F. Mean, top-level temperatures had been lowered 7°F in the 14 years since comparable 1952 studies. The probability curve for distribution times indicates that, for frozen products in general, there is a 5 percent probability that the product reach the consumer in 2 months or less, 50 percent probability that the time is 5 months or less, and 94 percent probability that the time is 10 months or less.

The most significant finding from all of these time-temperature surveys was that both times and temperatures are spread over wide ranges. Conditions in the top 10 percent were severe enough to cause deterioration in the quality of frozen foods.

Vitamin and mineral retention Freezing, per se, does not injure vitamins. It is mishandling after freezing which lowers the vitamin content. Ascorbic acid is frequently used as an index to poor packaging or mishandling since it is more readily lost than the B vitamins. Fig 15, taken from the fourth paper in the USDA time-temperature tolerance series,[28] illustrates the excellent retention of ascorbic acid in frozen peas during one year's storage at 0°F and the progressively more rapid and significant losses as the storage temperature was raised to 10°, 20°, or 30°F. At what might be considered a normal storage temperature of minus 4°F, losses of ascorbic acid from frozen food approximate 5 to 15 percent during 12 months.

The thirteenth paper in the USDA series[29] reported on studies using commercially frozen strawberries and raspberries stored at constant and fluctuating temperatures. Constant temperatures were 0°, 10°, and 20°F, while fluctuating temperatures during 24-hour or 72-hour cycles ranged from 5° to 10°F above and below the constant temperatures. Ascorbic acid retentions were the same for the constant and the comparable fluctuating conditions. Under somewhat comparable, but more extreme temperature fluctuation conditions, one-year storage studies with frozen concentrated juice[30] showed ascorbic acid to be extremely stable, with orange juice losing flavor or other qualities before it lost ascorbic acid. Bender[31] has tabulated ascorbic acid retention for fresh peas and peas processed by freezing, canning, air-drying, and freeze-drying. When they were prepared for dinner, he reported 44 percent retention for fresh peas; 39 percent for frozen; 36 percent for canned; 25 percent for air-dried; and 35 percent for freeze-dried.

Few studies appear to have been made on the retention of other vitamins. Carotene retentions in frozen vegetables are reported to be in the range of 80-100 percent during 1 year storage at about minus 4°F. The B vitamins appear to be retained to a high degree if the frozen vegetables are stored under conditions acceptable for conserving ascorbic acid, with quantitative retentions showing variation for different vegetables. Estimates of B vitamin retentions following 1 year storage at 0°F are 90 percent for asparagus, 75 percent for broccoli; 70 percent for green beans; 50 percent for cauliflower; 90 percent for peas; and 50 percent for spinach. The source does not make it clear whether these are overall retentions or retentions during storage.

Proteins, fats, and carbohydrates do not change nutritionally during normal storage of frozen foods. The same holds true for mineral content.

Packaging While there are many essential requirements for the packaging of various frozen foods, only oxygen permeability and light transmission have a direct bearing on nutrient retention. It is generally felt that hermetic seals and materials having limited or protective light transmitting characteristics are not required if storage and distribution is at 0°F or lower.

Measures to improve vitamin retention Voluntary and official standards of good practice for the frozen food industry are described by Sawyer and Hayes.[27] All of these are designed to safeguard the nutritional values and quality of frozen foods during warehousing and distribution by specifying maximum temperatures which should be used. In general, the standards require maintenance of 0°F or lower at all times, although the seven states which have promulgated regulations have tolerances ranging from 5° to 10°F for unavoidable variations.

FORTIFICATION

It is our opinion that foods of significant nutritional quality are now available in sufficient supply that adequate nutrition can be obtained by eating a suitable variety. We do not believe that processed fruits and vegetables should be made into vitamin pills, so that each would supply the RDA for all, or most, of the vitamins.

However, there are precedents for standardizing the ascorbic acid content of fruits and fruit drinks which are considered by the average consumer to be excellent sources of vitamin C. Such standardization by means of fortification or restoration should provide for variations in natural content and for losses which may occur during normal storage and distribution. Under present Standards of Identity, Code of Federal Regulations, Title 21 (21CFR), canned prune juice may be fortified with 30 to 50 mg of ascorbic acid per 6 fluid ounces; canned pineapple juice with 30 to 60 mg per 4 fluid ounces; and canned applesauce with 30 to 60 mg per 4 ounces avoirdupois. Additionally, proposals for Standards of Identity for juice drinks, fruit nectars, and juice cocktails request optional addition of 30 to 60 mg of ascorbic acid per 4 or 6 fluid ounces.

In accordance with the above philosophy, the National Canners Association is presently preparing a petition to amend the Standard of Identity for canned tomato juice to permit optional addition of ascorbic acid in a quantity sufficient to provide 10 mg per fl oz of the finished product. Optional labeling statements would be such as to avoid the "numbers game."

SUMMARY

The vitamin losses during storage and distribution of canned and frozen foods vary in magnitude but are not highly significant, unless the temperatures are well above normal or the time unusually prolonged. There is no evidence of mineral losses during storage and distribution. The biological availability of vitamins and minerals is not altered by storage. However, analytical methods used in determining vitamin and mineral content should be such as to reveal their biological availability.

Fruits and vegetables, whether fresh, frozen, canned, air-dried, or freeze-dried, show approximately the same percentage retention of ascorbic acid when finally prepared for consumption. Those which are significant sources of particular nutrients remain so after proper processing, storage, and distribution.

Table 33
Carotene Retention in Canned Tomato Juice and Peaches.

	Percent Carotene Retention					
Storage	Tomato Juice			Peaches		
Months	50°F	65°F	80°F	50°F	65°F	80°F
12	98	100	99	95	90	86
24	94	97	98	75	64	63

Table 34.
Carotene Retention in Canned Carrots and Yellow Corn.

	Percent Carotene Retention					
Storage	Canned Carrots			Yellow Corn		
Months	50°F	65°F	80°F	50°F	65°F	80°F
12	94	97	93	85	87	84
24	90	95	91	69	72	87

Table 35.
Riboflavin Retention in Canned Asparagus and Peas.

	Percent Riboflavin Retention					
Storage	Asparagus			Sweet Peas		
Months	50°F	65°F	80°F	50°F	65°F	80°F
12	93	87	83	93	89	84
24	81	77	72	88	84	81

Figure 14.
Retention of Ascorbic Acid During Storage of Canned Tomato Juice.

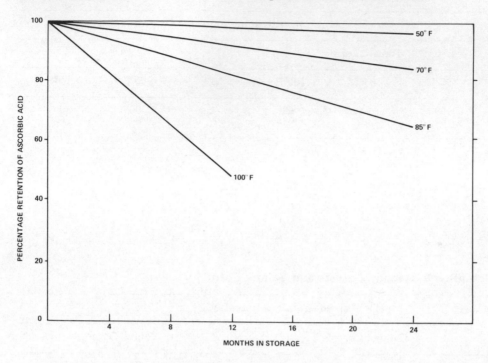

Figure 15.
Loss of Reduced Ascorbic Acid in Frozen Peas.

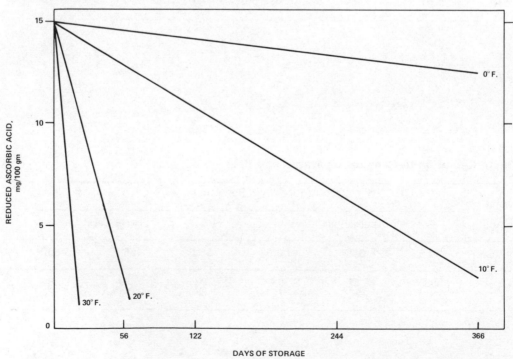

1. Cecil, S. R. and Woodroof, J. G. "Long-Term Storage of Military Rations." *Techn. Bull. N. S. 25*, Georgia Agr. Expt. Sta., Experiment, Ga., 1962.

2. Woodroof, J. G. and Lebedeff, O. K. "Foods for Shelter Storage—A Literature Review for the Office of Civil and Defense Mobilization." *CDM-SR-59-31, Rpt. No. 6*, Georgia Agr. Expt. Sta., Experiment, Ga., 1960.

3. Cecil, S. R. and Woodroof, J. G. "The Stability of Canned Foods in Long-Term Storage." *Food Techn. 18, No. 5*, 131-138, 1963.

4. Brenner, S. et al. "Effect of High Temperature Storage on the Retention of Nutrients in Canned Foods." *Food Techn. 2, No. 3*, 207-221, 1948.

5. Kramer, A. "Correlation of Quality and Nutritive Factors." *Food Packer*, April, 1948.

6. "Food Consumption, Prices, Expenditures." Supplement for 1968 to *Agricultural Economic Report No. 138*, USDA, Economic Research Service, 1970.

7. Anon. "Nutritive Value of Canned Foods (Part I and Part II)." *Nutrition Revs., 7, No. 5*, 142-146, 1949.

8. National Canners Association, Washington, D.C. "Canned Foods in Human Nutrition." 1950. (Out of print).

9. National Canners Association, Washington, D.C. "Retention of Nutrients during Canning." 1955.

10. Wagner, J. R., et al. "Effect of Commercial Canning and Short-Term Storage on Ascorbic Acid Content of Grapefruit Juice." *Food Res., 10, No. 6*, 469-475, 1945.

11. Guerrant, N. B., et al. "Influence of Temperature and Time of Storage on Vitamin Contents" *Ind. & Eng. Chem., 37, No. 12*, 1240-1243, 1945.

12. Lamb, F. C. "Factors Affecting Ascorbic Acid Content of Canned Grapefruit and Orange Juices." *Ind. & Eng. Chem., 38, No. 8*, 860-864, 1946.

13. Moschette, Dorothy S., et al. "Effect of Time and Temperature of Storage on Vitamin Content of Commercially Canned Fruits and Fruit Juices (Stored 12 Months)." *Ind. & Eng. Chem., 39, No. 8*, 994-999, 1947.

14. Sheft, Bernice B., et al. "Effect of Time and Temperature of Storage on Vitamin Content of Commercially Canned Fruits and Fruit Juices (Stored 18 and 24 Months)." *Ind. & Eng. Chem., 41, No. 1*, 144-145, 1949.

15. Guerrant, N. B., et al. "Influence of Temperature and Time of Storage on Vitamin Content." *Ind. & Eng., Chem., 40, No. 12*, 2258-2263, 1948.

16. Monroe, K. H. et al. "Some Studies of Commercial Warehouse Temperatures with Reference to the Stability of Vitamins in Canned Foods." *Food Techn., 3, No. 9*, 292-299, 1949.

17. Lamb, F. C., et al. "The Effect of Storage on the Ascorbic Acid Content of Canned Tomato Juice and Tomato Paste." *Food Techn. 5, No. 269-275*, 1951.

18. Harris, R. S. and Von Loesecke, H. "Nutritional Evaluation of Food Processing." John Wiley & Sons. Inc., New York, 1960.

19. Panalaks, T. and Murray, T. K. "The Effect of Processing on the Contents of Carotene Isomers in Vegetables and Peaches." *J. Inst. Can. Techn. Aliment. 3, No. 4*, 145-151, 1970.

20. Watt, B. K. and Merrill, A. L. "Composition of Foods, *Agriculture Handbook No. 8.*" Agr. Res. Service, USDA, Washington, D.C., 1963.

21. Rice, E. E. and Robinson, H. E. "Nutritive Value of Canned and Dehydrated Meat and Meat Products." *Amer. J. Public Health, 34, No. 6*, 587-592, 1944.

22. Bongolan, D. C., et al. "Effects of Process Technique, Storage Time, and Temperature on Thiamine Content of Plum-Tapioca and Split Peas with Ham." *Food Techn., 19, No. 8*, 83-85, 1965.

23. Theriault, F. R. and Fellers, C. R. "Effect of Freezing and of Canning in Glass and in Tin on Available Iron Content of Foods." *Food Res., 7, No. 6*, 503-508, 1942.

24. Reister, D. W., et al. "Temperature Variations in Warehousing Citrus Juice." *Food Ind., 20, No. 3*, 102-105, 224, 226, 1948.

25. Murray, R. V. "Storage of Canned Foods." *Bull. No. 17*, Continental Can Co., Inc., 1949.

26. Van Arsdel, W. B. "The Time-Temperature Tolerance of Frozen Foods. 1. Introduction—The Problem and the Attack." *Food Techn. 11, No. 1*, 28-33, 1957.

27. Van Arsdel, W. B., et al. *Quality and Stability of Frozen Foods*, Wiley-Interscience Division, John Wiley & Son, New York, 384 pp, 1969.

28. Dietrich, W. C. et al. "IV. Objective Tests to Measure Adverse Changes in Frozen Vegetables." *Food Techn. 11, No. 2*, 109-133, 1957.

29. Guadagni, D. C. and Nimmo, C. C. "XIII. Effect of Regularly Fluctuating Temperatures in Retail Packages of Frozen Strawberries and Raspberries." *Food Techn. 12, No. 6*, 306-310, 1958.

30. McColloch, R. J., et al. "VII. Frozen Concentrated Orange Juice." *Food Techn. 11, No. 8*, 444-449, 1957.

31. Bender, A. E. "Nutritional Effects of Food Processing." *J. of Food Techn., 1, No. 4*, 261-289, 1966.

The problems associated in the analytical division of any laboratory are directly proportional to the information background of the sample to be analyzed. With the advent of increased fortification of foods, this problem becomes greatly magnified. Knowledge of the complete food technology of the processed foods submitted for analysis allows for increased accuracy and precision. A general understanding of food consumption, calorie content, and the percentage that these foods play in a daily dietary is important to the analyst in his understanding of the analytical difficulties to expect.

Using the National Food situation figure as published by the USDA, 1969, we can see (Table 36) that foods can be compiled into six major contributors, each group making up a different analytical situation (1) meats, fish, and poultry products; (2) milk products; (3) fats and oils; (4) fruits and vegetables; (5) cereals; and (6) sugars. These foods total approximately 3.17 pounds per person per day and calculate 3,293 calories per person per day.

The percentage calorie contribution of these products (Table 37) is as follows: meat, fish, poultry 24 percent; milk products 11 percent; fats and oils 18 percent; fruits and vegetables 12 percent; cereals 18 percent; and sugars 17 percent. Analytical problems in processed foods at the moment are confined to three classes—milk products, fruits, and vegetables, and cereals. Sugars may be expected as a fourth class as fortification guidelines are developed. Other classes may be added as food fortification guidelines are developed. Textured vegetable protein may offer some new problems.

The vitamin, mineral and calorie levels in the Recommended Daily Dietary Allowances (RDDA) 1968 for a male subject 14 to 18 years of age are given in Table 38. The recommended daily dietary allowance of 3,000 calories closely approximates the United States Department of Agriculture (USDA) calorie consumption figures. The assumption is also made that future fortification programs may be based on a units per gm of food consumed, 1,500 gm per day, or on a units per calorie basis. Table 38 is set up on a fortification per gm basis or 1,500 gm of food consumed per day. It illustrates the analytical detection level that can be obtained using present day methodology.

Very clearly it can be noted that

chapter 7
analytical problems associated with determining content and biological availability of vitamins and minerals in processed foods

by
Philip H. Derse, MS

vitamin D is the only nutrient where difficulty occurs. Chemical vitamin D methodology is not available for across the board food fortification and the rat bioassay procedure must be used, which is costly and time-consuming. However, on the basis of Table 38 it is clear that with adequate sample information submitted to the laboratory it can perform the necessary analytical work. A compilation of current methods employed for analyzing vitamins and minerals appears at the end of this paper.

In order to understand a few of the problems encountered in the analytical area, each vitamin or mineral will be briefly discussed. The majority of these difficulties are not found in the methodology and must be obtained through experience.

VITAMIN A

Difficulties occur at low levels and in general require chromatography to separate the vitamin A from other materials. High fat materials cause trouble because of necessary increased saponification. Sterols will still remain to interfere with calorimetry procedures.

United States Pharmacopeia (USP) spectrophotometric vitamin A analysis does not take into consideration the presence of vitamin E. Vitamin E, because of overlapping absorbances of the ultraviolet curves, will cause serious error. Use of stabilized vitamin A concentrates must be recognized in order to employ correct extraction procedures. Though biological vitamin A analysis with rats is still employed, there appears to be little difference between correct chemical analysis and biological assay.

VITAMIN D

Vitamin D chemical analyses are always questionable. Presence of tachysterol in D resins at 5 to 15 percent interferes with the analysis. This is especially true of materials in accelerated storage.

Despite the fact that biological rat assays can be performed on low potency materials, the influence of the accompanying food nutrients, such as calcium and phosphorus, may interfere with the accuracy of the analysis. Table 39 reviews several of the variables in analysis that occur. Biological analyses employing the rat or chick vary greatly from the chemical analyses available today.

VITAMIN E

The greatest error in the chemical analysis for vitamin E is the fortification material employed. The assay is unable to differentiate *d* and *dl* alpha tocopherol; thus, no unit value can be recorded. For example:

1 mg dl alpha tocopherol = 1.1 unit vitamin E

1 mg d alpha tocopherol = 1.49 unit vitamin E

As in most fat soluble vitamins, antioxidants cause serious interference and must be removed to avoid high results. Absorption of the vitamin in the food, either directly or in cases of protein binding in heated products, will result in low analysis unless appropriate extraction steps are taken. Again, though biological rat assays are employed, they are tedious and costly. Good agreement between chemical and bioassay can be expected.

ASCORBIC ACID

In processing various products, reduced iron salts are sometimes added which cause high results in the assay. Other trace elements such as copper, manganese, and iodine often added to products cause rapid losses of ascorbic acid during extraction. During heat treatment such as baking, ascorbic acid may be bound to either protein or may form a carbohydrate complex which will not extract or will not react in the normal way in the assay.

Heat treated products often give erroneously high results due to reductones and other reducing substances that are generated during processing. Certain products are treated with sulfites during processing (dehydrated potatoes and other). Sulfite causes high results unless it is eliminated by chemical treatment. Recently iso-ascorbic acid is being used as a food preservative. Current methodology will not differentiate ascorbic acid and iso-ascorbic acid.

THIAMINE

In the formulation of certain products, ascorbic acid added in sufficient excess will result in a low negative thiamine content. The ascorbic acid prevents the development of thiochrome. Defoamers, stabilizers, and emulsifiers used in dehydrating result in high fluorescent blanks and cause variable and inaccurate results. High levels of various iron salts react with thiamine during extraction and give low results in the assay. Specific treatment is necessary to obtain complete recovery of products fortified with thiamine phosphate salts. These salts do not respond in the assay until the salts are hydrolyzed enzymatically.

Extraction problems are encountered with thiamine as with ascorbic acid when they are in a wax base. Hot acid extraction in an autoclave will not give a quantitative extraction with most products. Folacin, niacin, riboflavin, vitamin B_6 and vitamin B_{12} are analyzed employing microbiological techniques. Problems of extraction, food additives, protein, all play important roles in the accuracy of the result obtained.

As the response of the bacteria or yeast growth involved in the technique are compared to their growth in standard media, it becomes immediately evident when inhibition of growth or enhancement of growth occurs. Additional extraction procedures must be developed to obtain accurate results. Color of the sample and precipitation during bacterial acid production also add to the difficulty.

In the vitamin B_6 assay, no attempt is made to differentiate between the three B_6 forms. Frequently, media employed are contaminated with the vitamins that are to be analyzed. When pseudo forms of B_{12} are suspected, additional work is involved employing the organism Ochromonas. However, this latter organism does not allow for the sensitivity necessary in food fortification. Biological assays involving rats and chicks have been studied extensively and only in the case of pseudo vitamin B_{12} is there ever any reason to doubt the microbiological analysis.

There are very few problems in the chemical analysis of food products for minerals. This is due to the fact that most mineral analyses are conducted on a wet (acid) or dry ashed sample, which gives you a sample relatively free of interfering materials. This sample is also homogeneous at this point and methods used are specific for a certain element. Most problems in food analysis come from non-homogeneous samples and also insufficient blending or mixing of samples at the time the analysis is initiated.

Atomic absorption methods give very good results due to the specific nature of the instrumentation. Problems are nearly always due to incorrect claims, poor mixing of original sample, or poor preparation. In determination of iodine, most problems are caused by incorrect claims and improper mixing or storage. With the right information and handling, no real problems are presented. Phosphorus uses an ashing technique and the colorimetric method is accurate and very specific. No problems are known.

Emission spectrometry (direct reading) of high levels of some elements will give high results on others due to the overlapping of high intensity lines.

However, this occurs with only a very few samples and can be corrected for if proper information is available.

Bioassays are used primarily as check assays and as a basis for standardization of the more rapid physio-chemical methods. The chief advantage of animal assays is that they are based on biological response, which is important from the nutritional standpoint. Chemical assays can give the level of a mineral or vitamin present in a food stuff, but cannot tell anything about the availability. The disadvantage of bioassays are that they are time-consuming, expensive, and leave something to be desired insofar as precision is concerned in many instances.

The type or kind of food in which the mineral or vitamin is being measured may present problems. Animals may object to it or not tolerate it well. A high mineral or salt content may cause extreme diarrhea and interfere with an assay. A high fat or high moisture content may prevent sufficient intake of nutrients to show response. In these instances, fat extractions may be required. High moisture may be removed by freeze-drying or other suitable means which do not alter the item being measured.

In some instances, the level of the mineral or vitamin may be so low that the animal cannot consume enough food to show a response. It may be necessary to concentrate by extraction or other means. If this is not possible, it may be impossible to assay properly. This is especially true with many of the vitamins.

Bioassays are valuable in determining mineral availability and must be taken into consideration when accurate iron, calcium, phosphorus, and magnesium information is desired. For example, the greater part of phosphorus in plants is phytin phosphorus which is almost unavailable. This may constitute as high a level as 60 percent of the chemical phosphorus. Many foods contain moderate amounts of oxalic acid, tying up portions of the calcium and magnesium in the products. The chemical analysis thus loses much of its significance.

As an example of the difficulty described, a review of a collaborative study conducted by Pla and Fritz, 1970, is illustrated in Table 40. Twelve forms of iron salts and eight foods, three fortified, were analyzed on the basis of their chemical iron content for biological availability. Values range from 2 to over 100 when compared to iron sulfate as 100.

Processing of food may change the form of iron employed as a supplement; thus, availability studies must be conducted. Our largest errors in assays of calcium, magnesium, iron, and, possibly, manganese, arise from the fact that we do not take the time to run through a biological availability test. The fact is that dried egg yolk gives you a good source of iron, but the iron is only 33 percent available. We can add ferrous sulfate and increase the effect by the amount of iron sulfate we add.

In review, the validity of chemical, microbiological, and bioassay analyses of processed foods can be greatly enhanced through knowledge of the ingredients used in the food and the type of process employed. Initial bioassays must be conducted in processed or fortified foods to accurately evaluate the availability of calcium, phosphorus, and iron. Following availability studies chemical procedures may then be employed for calcium, phosphorus, and iron.

BIOLOGICAL ASSAY METHODS

Vitamin A—Chick: Foy & Morgareidge: "Liver Storage Tests of Guggenheim & Koch." *Anal. Chem.* 20: 304, 1948.
Vitamin A—Rat: *Pharmacopoeia of the United States*, 13th Revision, 719, 1947.
Vitamin C—G. Pig: Sherman, et al: "Quantitative Determination of Antiscorbutic Vitamin." *Jour. Am. Chem. Society*, 44: 165, 1922.
Vitamin D—Chick: *Methods of Analysis of the A.O.A.C.*, 10th Ed., 784, 1965.
Vitamin D—Rat: *USP*, XV: 889, 1955.
Vitamin E—Rat: Mason & Harris: Bioassay of Vitamin E." *Biological Symposia*, XII: Jacques Catell Press, Lancaster, Pa. 1947.
Iron—Rat & Chick: Pla & Fritz: "Availability of Iron." *Jour. of the A.O.A.C.*, 53, 4; 791, 1970.
Phosphorus and Calcium: Dilworth & Day: "Phosphorus Availability Studies with Feed Grade Phosphates." *Poultry Science*, 43: 1039, 1964.

MICROBIOLOGICAL ASSAY METHODS

Niacin: *A.O.A.C.*, 11th Ed.: 787, 1970.
Riboflavin: *A.O.A.C.*, 11th Ed.: 789, 1970.
Vitamin B_6: Atkins, Schultz, Williams, and Frey: *Ind. & Eng. Chem.*, Anal Ed., 15: 141, 1943.
Folacin: *A.O.A.C.*, 8th Ed.: 830, 1955.
Vitamin B_{12}: *USP*, XVII: 864, 1965.

CHEMICAL PROCEDURES

Vitamin A; *USP, XVII*: 890, 1965; *A.O.A.C.*, 10th Ed.: 755, 1965; *J. Dairy Science*, 31: 315, 1948.
Vitamin E: *Acta Chemica Scandinavia*, 11: 34, 1957.
Ascorbic acid: *USP*, XVII: 48, 1965, *J. Biol. Chem.*, 160: 217, 1945; *J. Biol. Chem.*, 147: 399, 1943.
Thiamine: *USP*, XVII: 888, 1965, *A.O.A.C.*, 10th Ed.: 758, 1965.
Riboflavin: *USP*, XVII: 886, 1965; *A.O.A.C.*, 10th Ed.: 762, 1965.
Vitamin D: *USP*, XVII: 891, 1965.
Iodine: *A.O.A.C.*, 11th Ed.: 674, 1970; W.T. Binnerts, *Anal. Chem. Acta*, 10: 78, 1954.
Phosphorus: Fiske & Subbarow: *J. Biol. Chem.*, 66, 375, 1925; *A.O.A.C.*, 11th Ed., 11 1970.
Calcium and Iron: *Analytical Methods for Atomic Absorption Spectrophotometry*. Perkin-Elmer.

Table 36.
National Food Consumption Calorie Contribution.

Food	Pound per individual per year	Calories per pound	Calories consumed per year
Meats	182.0	1200	218400
Fish	11.1	170	1870
Chicken	39.0	300	11700
Turkey	8.4	600	50400
Eggs	35.0	600	21000
Cheese	10.9	1500	16350
Cond. & evap. milk	8.0	600	4800
Fluid milk & cream	271.0	300	71300
Ice cream	18.1	900	16200
Fats and oils	53.0	4000	212000
Fruits, fresh	79.1	300	23700
Fruits, canned	24.4	300	7320
Fruit juice, canned	14.6	200	2920
Fruit juice, frozen	9.3	700	6510
Fruit juice, dried	2.6	1700	4420
Fresh vegetables	97.9	300	29370
Canned vegetables	51.1	295	15045
Frozen vegetables	9.1	295	2655
Potatoes	118.2	275	32450
Sweet potatoes	6.0	400	2400
Beans, dry	6.8	1550	10540
Melons	23.1	65	1495
Corn syrup	15.0	1315	19725
Corn sugar	4.8	1600	7680
Sugar	99.9	1750	174825
Corn meal	7.4	1600	11840
Wheat flour	112.0	1600	179200
Wheat cereals	2.9	1600	4800
Rice	8.3	1600	13280
Coffee	14.1	600	8400
Cocoa	4.0	1200	4800
Total per year	1353 pounds		1202145 calories
Total per day	3.71 pounds		3293

Table 37.
Percent Distribution of Calories.

Meats, fish, poultry eggs	24%
Milk and milk products	11%
Fats and oils	18%
Fruits and vegetables	12%
Cereals	18%
Sugar	17%

Table 38.
Fortification Levels and Analytical Detection Limits Recommended Daily Dietary Allowances—1968.

	Male 14-18 years	Fortification level per gm food	Analytical detection level	Method
Calories	3000.0			
Vitamin A	5000.0 IU	3.300	1.000 IU	Chemical
Vitamin D	400.0 IU	0.280 IU	0.200 IU	Rat bioassay
			1000.000 IU	Chemical
Vitamin E	25.0 IU	0.018 IU	0.005 IU	Chemical
Ascorbic acid	55.0 mg	38.000 y	10.000 y	Chemical
Folacin	0.4 mg	280.000 my	8.000 my	Microbiological
Niacin	20.0 mg	13.000 y	1.000 y	Microbiological
Riboflavin	1.5 mg	1.000 y	0.200 y	Microbiological
Thiamine	1.5 mg	1.000 y	1.000 y	Chemical
Vitamin B-6	1.8 mg	1.2000 y	0.040 y	Microbiological
Vitamin B-12	5.0 y	3.300 my	0.080 my	Microbiological
Calcium	1.4 gm	980.000 y	0.500 y	Spectrometer
Phosphorus	1.4 gm	980.000 y	1.000 y	Spectrometer
Iodine	150.0 y	0.100 y	0.050 y	Chemical
Iron	18.0 mg	12.00 y	1.000 y	Spectrometer
Magnesium	400.0 mg	280.000 y	0.100 y	Spectrometer

Table 39.
Vitamin D Analysis—Chemical vs Biological.

Sample	Claim	Chemical	Biological
Dry chocolate beverage powder	450 u/100 gm	760 u/100 gm	(R) 450 u/100 gm
Vit. D capsules	1,500 u/cap	1,500 u/cap	(R) 1,590 u/cap
Vit. D capsules	1,000 u/cap	1,300 u/cap	(R)-1,500 u/cap
Vit. D in oil	3,500,000 u/gm	3,430,000 u/gm	Rat: 4,900,000 u/gm
			Chick: 4,500,000 u/gr
D_3 resin	28,000,000 u/gm	29,300,000 u/gm	(R) 23,400,000 u/gm
Irradiated	200,000 u/gm	278,000 u/gm	(R) 208,000 u/gm
Vit. capsules	50,000 u/cap	54,500 u/cap	(R) 69,500 u/cap

Table 40.
Relative Biological Value of Iron (FeSO$_4$ = 100). [*]

Ferrous sulfate	100
Ferric ammonium citrate	107
Ferric choline citrate	102
Ferrous chloride	106
Ferrous gluconate	97
Ferric chloride	78
Ferric pyrophosphate	45
Reduced iron (electrolytic)	46
Reduced iron (by hydrogen)	25
Ferric orthophosphate	15
Sodium iron pyrophosphate	12
Ferrous carbonate	2
Breakfast cereal (with reduced iron)	43
Enriched flour (with reduced iron)	32
Corn germ meal	40
Oat flour	21
Wheat germ meal	54
Blood meal	35
Dried egg yolk	33
Egg white + ferrous sulfate	79

[*]Pla and Fritz: "Availability of Iron." *Journ. of A.O.S.C.*, 53: 791, 1970.

Q: We haven't said much about newer foods, or old foods whose new forms have become more predominant. There was a brief discussion of the opportunity of improving white potatoes by breeding and doubling the ascorbic acid content, but nothing said about the fact that somewhere between 16 and 18 percent of all potatoes harvested in the United States end up as dehydrated potatoes. Dehydrated potatoes which are prepared in an institution have zero vitamin C content. Whipping them in a large operation destroys the C consistently. The American Potato Growers Association is aware of this and has wanted to put vitamin C back into potatoes, but there was some disinterest in that on the part of the United States Food and Drug Adminsitration (FDA).

Another problem with vitamin C is the introduction of paper milk cartons for orange juice. Fifteen percent of all orange juice is marketed in paper milk cartons. Don't assume you have 50 mg of ascorbic acid per 100 ml despite what your physician tells you. You may have zero. In no case will you have anything more than two-thirds of what *Handbook 8* says you should have. These are important changes that we in food technology are responsible for, sometimes unawaredly so.

We need data on other vitamins. We've discussed four vitamins. Nothing was said about folic acid; nothing was said about E or B_6; very little about B_{12}; and a bit about pantothenic acid.

There's almost nothing in the literature about the stability of folic acid in food processing. Those who know say we don't even have a good analytical method to determine whether free folate versus different confugates are present, and which ones are utilizable by man. There is one published reference to an alcoholic in Boston who had only one source of folic acid, hamburges from a restaurant. Victor Herbert went to the restaurant and assayed the hamburgers. They were kept on a steam table for hours and they had zero folic acid. Similarly, boiling milk destroys

free folic acid 100 percent. Pasteurization doesn't.

Potato chips are an excellent source of vitamin C, but some people eat dehydrated potato chips, that is, simulated chips made by taking dehydrated potatoes and reassembling them. That product has no vitamin C.

Vitamin E is unstable during freezing. We need much more information about this before we recommend that someone only eat certain vegetable oils because they're the only ones with high E content. It may not be there when we want it.

A (Dr. Barnes): Ascorbic acid is one of the nutrients that has shown up in various food consumption surveys to be at borderline to low intake. Of course, calculations of nutrient intake are based upon conversions from values given in *Handbook 8*. We have been asking some individuals quite knowledgeable in dietetics and nutrition about the source of ascorbic acid in the American diet, and we were surprised to learn that the new vegetable-potato type of mean is not considered to be a food that is expected to provide ascorbic acid intake.

But if we go back to the USDA data showing availability of nutrients to the population, 38 percent came from vegetables. Something like 20 to 25 percent of available ascorbic acid consumed in the United States was from the vegetable source, and about 20 percent from potatoes. The total adds up to at least 50 percent. To amplify what was said about potatoes, not only is there zero value in dehydrated potatoes, but also low value in other types of processed potatoes. We wonder if anyone knows of any values that are realistic in terms of the ascorbic acid intake of the American public. It would appear to be something very, very much less than the values that have been calculated from food consumption surveys.

Q: We ought to think about vitamin K in new foods. How about vitamin K and the question of bioassays versus chemical assays?

A (Dr. Sarett): The biological assay has been the only acceptable assay until now. We have a good chemical assay which people in our research laboratory have developed. Unfortunately, it is not an easy assay, and it has been applied mainly to infant formulas thus far.

On the question of iron availability, we studied the availability of various iron salts such as were listed in Table 40 in Chapter 7. When these are added per se to rat diets, we found the same relationship in which ferrous sulfate is 100 percent active. Various other salts fall into line at various activities, and some of the salts that are used in foods are quite low. However, if we take these same salts and add them to the food before processing, such as we have done with soy and milk infant formulas, the availability of the iron frequently comes up markedly.

Instead of being 20 or 30 percent as active as ferrous sulfate, they are 70 or 80 percent of ferrous sulfate. We just can't take iron salts and compare them without having them interact in the food, in the process that is carried out in manufacture.

A (Dr. Darby): We mustn't lose sight of the fact that there are a legion of other things than nutrients in foods that are important to nutritionists or to the consumer—such things are goitrogens in a wide variety of foodstuffs. As one alters the characteristic of plants for improving nutrient content we think we must be careful that we don't simultaneously alter the characteristic in a detrimental way, by increasing the content of certain of the toxic materials which naturally occur in foods.

Q: In Chapter 6, we went back to the 20 to 30-year-old studies of the National Canners Association on vitamins in canned foods, but since that time there have been some really tremendous changes in the canning practices. Can we say what the situation is today compared to 1942?

A (Dr. Farrow): Well, we're looking into this right now, as a matter of fact. There is a tremendous body of data on the nutritive value of canned foods, determined some 25 to 30 years ago. We have launched a program to review some of this on a pilot scale. We've completed one product so far, tomato juice. The indications are that no really big changes have occurred in the vitamin C content of tomato juice available today on the market

as compared with that sampled some 20 years ago. The number of samples were about the same. The sampling was balanced geographically to be roughly in proportion with the producing areas. The average vitamin C content in the current study was 13.4 mg per 100 gm and in 1950 the result was 14.4 mg.

Q: Going back to the question of iron availability, we ask if the rat bioassay procedure does in fact correspond for different types of foods and different forms of iron with absorption by man as measured by studies using radioactive iron in the human diet?

A (Dr. Hein): We don't know. We have performed comparable studies on dogs, chickens, and pigs but we don't really know of any strenuous study in man on these various products.

Q: In the comment regarding the reduction of vitamin C in tomatoes through processing, was Chapter 5 referring to mechanical harvesting?

A (Dr. Hein): No, it was referring to the suggestion which arose a few years ago to two things that happened to tomatoes. First the pH of the freshly harvested tomato had gone up with the change in variety. Secondly, the ascorbic acid variations in processed products was larger than previously and, in fact, in many cases may have fallen in some products. The question arose as to whether this was actually due to a change to the mechanical harvest variety or was due to a change in processing conditions, or possibly to isolated processes.

We can only answer in part for the mechanical harvesting. It has been stated that the varieties used in mechanical harvesting have a lower vitamin C content and we've looked into that quite thoroughly. We have studied our records over the years before and after mechanical harvesting and find that if there is any change, the vitamin C values of the mechanically harvested tomato are equal to or a bit higher than older varieties. We understand that Bernie Schweigert has been doing some work along the same line in California, as has Olaf Mickelsen at Michigan State.

Q: There is considerable discussion today regarding fortification or restoration of certain nutrients lost during processing, or adding of nutrients that were present at a very low level. We have to consider whether it actually can be done.

We have looked into the possibility of adding nutrients, including iron, copper, calcium, methionine, and vitamin A; and it's not a very easy thing to do.

The iron form is a very important factor. Also, to what do you add the iron, frozen food or dehydrated or pre-processed food? Iron changes the flavor in many cases. It may change the color, or it may, in some instances, alter the texture. Copper, which causes oxidation, may give a rancid fat, or a loss of vitamin C. Calcium is also a difficult nutrient to add. It can impart peculiar texture to products, and an odd flavor. Methionine, particularly in heat-processed products, develops a very distinct off-flavor in a food.

A (Dr. Sarett): Studies in animals in which ferrous sulfate is presumed as 100 percent available don't necessarily mean that iron sulfate is 100 percent utilized. It means that this is the standard. When iron in different forms was added and baked in bread, it was found that most forms of iron were utilized to about the same extent, with not too much difference between ferrous sulfate and other forms. Now, if we take our data, in which we've processed different iron salts in infant formulas and fed them to animals, we come up with about this same type of finding, which is different from Chapter 7's Table 40 that describes addition of salts per se to the animal's diet. We feel that availability studies in a rat, when done properly and done with foods in which the iron is incorporated in the food process, can give you a pretty good picture of what's happening in man.

A (Dr. Chichester): We think we might add one thing that we haven't really discussed, the problem of measuring what is actually consumed. We've talked about processed food, in which losses may be minimal or significant. In many cases, we've contrasted this against fresh materials. We've measured the reduction that may be due to processing, storage, or handling. Obviously, fresh products as consumed are often processed also, and we wonder if sometimes we get confused between processed products which are usually eaten with very little cooking and those which are said to be consumed fresh, but actually undergo considerable cooking.

Also, we eat many of our meals outside of the home. Something on the order of 30 percent of the money spent for food is spent for food consumed outside of the home. We have a development of institutional feeding in which processed products or semi-processed products are heavily used. We don't know of any significant data on the analysis of food as eaten outside the home, and we can guess there must be losses in processing. We could be neglecting some of the largest processing losses in the food supply.

fortification
of processed foods

PART
THREE

The modern fortification of foods for nutritional and public health purposes is a comparatively recent development and one which has been most effective and successful. My purpose in looking at it historically is to point out lessons that can be learned which are applicable to today's problems.

Fortification originated with the use of iodine in table salt. This was followed by the fortification of margarine with vitamin A concentrates; the fortification of milk with vitamin D; the enrichment of white bread, flour, and corn meal with thiamine, riboflavin, niacin, and iron; and the addition of vitamin C to various beverages. Recently, the availability in commercial quantities of the amino acid lysine is opening up a new era in the fortification of processed foods by improving the biological value of the protein that they contain.

The discovery of iodine, its presence in the animal body, the successful treatment of goiter with iodine, and the discovery that iodine is universally distributed in foodstuffs and drinking water and is deficient in areas where goiter is endemic, all laid the groundwork for early studies in the use of iodine in the prevention of goiter. Boosingaut in 1833 appears to have been the first to suggest the iodization of salt as a method of preventing goiter and in 1849 Grange recommended one part in 10,000 of iodine in kitchen salt.

These recommendations were ignored and it was not until the work of David Marine and his associates that the present-day iodization of salt was introduced. The first studies were carried out by Marine and Kimball in Akron, Ohio, in 1919 and 1920. The success of these trials was followed by large scale prophylaxis with iodized salt in Michigan in 1924, with the result that in five years the goiter rate fell from 38.6 percent to 9 percent and no toxic effects were observed. The practice has now become generally widespread and there is no longer any question concerning the efficacy of this method of controlling goiter as a public health problem.

In spite of the success of this measure, it has never been possible to secure legislation requiring the iodization of all table salt in the United States, with the result that the use of iodized salt in this country is variable and a constant educa-

chapter 8
past experience in fortification of processed foods

by
William H. Sebrell, Jr., MD

tional campaign is required in order to hold down the incidence of goiter, despite the demonstration in many parts of the world that there are no toxic effects. Doubts and confusion spread by opponents of the idea have blocked all attempts at legislation even though supported by the American Medical Association, the National Research Council, and the American Public Health Association.

The United States now lags behind many countries in failing to enact and enforce this demonstrated public measure. Therefore, while iodized salt stands as our first and very successful effort at the fortification of processed foods, it also remains as an example of the difficulties of getting good health protective measures enacted into law and regulations when they apply to the food supply.

The fortification of margarine with vitamin A is the second example of a major nutritional advance through the fortification of a processed food. The modern work which led to the fortification of margarine begins with the studies of Bloch (published in 1917) working with infants near Copenhagen during the years 1912 to 1916. These studies by Bloch laid the groundwork for the fortification of margarine with fish oil concentrates containing vitamin A. However, it was not until 1939 that the Council on Foods and Nutrition of the American Medical Association approved the addition of vitamin A to margarine, and in 1941 the definition and standard of identity for oleomargarine was promulgated by the Food and Drug Administration.

In 1952 the Standard was amended to specify that margarine contain at least 15,000 units of vitamin A per pound. This regulation was designed to give the margarine a vitamin A content comparable to that of butter. The vitamin A content of butter may vary from values below 10,000 IU per pound in winter to better than 20,000 IU in summer butter. Average butter has a vitamin A activity (including carotene) equivalent to about 15,000 IU per pound. Although margarine is standardized at 15,000 IU per pound, efforts to have butter standardized to a similar value have failed.

The recent availability of a water-miscible form of vitamin A has greatly extended the possibility of the prevention of vitamin A deficiency. It is well known that dry skim milk fed to children suffering from severe protein calorie malnutrition may precipitate severe vitamin A deficiency. Here we have a situation where taking a natural food without fortification produces what may be called a toxic effect which is corrected by the fortification.

Skimmed milk without vitamin A fortification should never be given to seriously malnourished children. Water-miscible vitamin A now makes it possible to add vitamin A to dry skim milk, and thus make the product better for use in areas in which protein calorie malnutrition is prevalent and where it is not feasible to introduce either dry whole milk or a food fat fortified with vitamin A. It was not until 1965 that nonfat dry skim milk, fortified with vitamins A and D to the level of 5,000 IU of vitamin A and 500 IU of vitamin D per 100 gm was provided for export in connection with our Food for Peace program.

Our next nutritional fortification of processed food was the addition of vitamin D to dairy products. The addition of vitamin D to cow's milk was first achieved in the 1920s by feeding cod liver oil to the cow, and later by feeding irradiated yeast and irradiated ergosterol, and then by the direct irradiation of milk with ultra violet light. Vitamin D is now added directly to the milk. It is generally acknowledged that the addition of vitamin D to milk has been the major factor in the disappearance of rickets as a public health problem in this country.

This early example of the fortification of a processed food illustrates many of the problems involved in such a procedure. There was considerable discussion as to the quantity to be added and the form of vitamin D to be used. The American Medical Association's Council on Foods and Nutrition officially endorsed vitamin D, favoring a level of 400 IU per quart and, because so many babies were on evaporated milk formulas, the decision to add 400 units of vitamin D to all evaporated milk per reconstituted quart was a very significant public health measure in that it provided automatic protection against rickets for about three-fourths of all the artificially fed infants in the United States. Vitamin D fortified milk as a means of preventing rickets made it only natural that the fortification of milk with other vitamins and minerals would be proposed and promoted.

Multivitamin, mineral-fortified milk and milk products were introduced in the early 1940s in an attempt to supply in one quart of milk the adult requirements of the most important vitamins and minerals, adding thiamine, riboflavin, niacin, iron, and iodine, as well as vitamins A and D. This had not become a successful widespread procedure.

In the 1930s, it was evident that beriberi, pellagra, riboflavin deficiency, and iron deficiency anemia were widespread enough in this country to constitute public health problems. Considerable thought was given to the problem of adding the necessary nutrients to some food in order to prevent these conditions and eliminate them as public health problems. The necessary nutrients were available in commercial quantities, at prices which made the proposal economically feasible. It was obvious that the natural food carriers for these products were the cereal grains, wheat, corn, and rice, which are the mainstays of the American diet.

The concept of the addition of synthetic nutrients to foods was so widespread and acceptable as a health measure at that time that in 1939 the Council on Foods and Nutrition of the American Medical Association adopted a policy on the addition of vitamins and minerals to food, and in 1941 the Food and Nutrition Board of the National Research Council announced its policy on the addition of synthetic nutrients to foods. These Statements on Policy were jointly published in 1953 and revised in 1961 and 1968.

The addition of vitamins to white flour and bread was first proposed for official recognition in 1940. Hearings were held by the Food and Drug Administration to establish a definition for "enriched bread" which was first issued on May 27, 1941. The levels of the required ingredients of thiamine, riboflavin, niacin, and iron were based on clinical evidence of the prevalence of these deficiencies in this country, estimated shortages in the intake of these nutrients in the typical American diet, and assumptions as to the extent to which flour and bread entered into the diet.

Two schools of thought developed at that time about the addition of synthetic nutrients to foods. The philosophical question involved was, basically, whether these nutrients should be added to foods in quantities which restored the vitamin levels to those approximating high levels found in the natural food, or whether the amount of nutrient to be added should be based on estimates of the quantity necessary to prevent the disease, without regard to the amount naturally present in the food.

This question has never been entirely resolved. To me, it has always seemed that, since the addition of synthetic nutrients was for the purpose of preventing disease, the quantity used should be based on the amount necessary to be effective and safe, rather than the amount that happens naturally to be present in a foodstuff.

The enrichment of white bread and flour was based on public health objectives. The enrichment formula was designed to attain the important objective of improving the nutritive value of the diet without requiring changes in food habits, without altering the taste or appearance of the product, and using foods basic to practically every diet in this country. In addition, the material was low in cost and readily obtainable. These objectives are desirable in any plan to use synthetic nutrients.

If we look at the medical records of the 1930s, the low level of intake of thiamine, niacin, and riboflavin of a large part of the population resulted in the wide prevalence of the vitamin deficiency diseases (beriberi, pellagra, ariboflavinosis), while iron deficiency anemia was widespread then, as it still is today. The medical and public health professions were deeply concerned. The Food and Nutrition Board of the National Research Council gave it much study, and the US Department of Agriculture was actively trying to combat the problem. It was clear, however, that the situation was one which required more rapid action.

It was obviously impossible to make any rapid progress through educational methods; it was economically out of the question to distribute the necessary foods to the people who needed them the most; and it was impractical to get vitamin pills taken regularly by large numbers of people who needed them. Attempts to get people to eat whole grain products had been going on for years with little success. The problem, therefore, narrowed down to the question of how to provide the newly available synthetic nutrients to the general population in the simplest and most economical way.

The average consumption of bread at that time was calculated to be about six slices per day. It was a simple matter to determine, on the basis of the Recommended Allowances of the National Research Council, how much thiamine, riboflavin, iron, and niacin should be added to flour and bread so that six slices per day per person would bring the intake of nutrients of most people up to the Recommended Allowances and thereby raise the level of these nutrients in the diet high enough to protect against deficiency disease.

The job was accomplished by consultation with the people who were to be responsible for carrying out the program. The proposal was placed before a meeting of the official organizations of millers and bakers and members of the medical and public health professions, as well as nutritionists, government officials, and scientists. The proposal received immediate and wholehearted endorsement. An important factor was the determination that the proposed additives would not change the product in form, appearance, or flavor. The US Food and Drug Administration set the standards for enriched bread and flour with minimal and maximal levels of nutrients.

Following the introduction of enrichment, beriberi, ariboflavinosis, and pellagra disappeared as public health problems in this country. This has led to differences of opinion as to the role played by enrichment in eliminating these diseases.

This is a question that can never be completely and positively answered since it was impractical to set up an experiment to prove its value. I doubt very much that it is ever going to be possible to set up an experiment that will conclusively prove the benefits of the addition of any single synthetic nutrient to a food as applied to large population. Always occuring at the same time are other events, such as changes in the economic situation, changes in food intake that are not controllable experimentally, the occurrence of epidemics, and other factors that cannot be anticipated or controlled.

The enrichment program was carried out without any direct experimental evidence that the enrichment would be successful or effective. It seemed clear enough that, since it had been demonstrated that thiamine would prevent and cure beriberi, any food to which thiamine would be added would help prevent the disease. The same reasoning was applied to pellagra, and to ariboflavinosis. However, questions have continued to arise over the past 25 years as to whether the enrichment program was effective and people still seem to want direct proof which is practically impossible to obtain.

In any procedure of this kind, affecting the general population, it is exceedingly difficult to design a study that can ethically be carried out, to prove that the product is effective. If it is done on a confined selected group of individuals, the study is criticized on the basis that these conditions do not apply to the general public. If a large-scale study is made of the general public, such as the population of the United States over 25 years, during which these diseases have ceased to exist as public health problems, one is faced with the situation that many other things occurred at the same time and the effects of the enrichment program cannot be separated from those of nutrition education, food availability, food prices, income, etc.

Even when the US Department of Agriculture shows by its family dietary studies that the vitamins present in enriched bread and flour make a substantial contribution toward having a deficient diet reach recommended allowances, the argument is stated that these data do not apply to individuals. The statistics indicate that the result was beneficial and we know that thiamine, niacin, and riboflavin deficiency have disappeared as public health problems.

Vitamin C is one important synthetic nutrient which had been neglected in our program—probably because vitamin C deficiency was not recognized as a public health problem. We should have long ago standardized and regulated the vitamin C content of citrus and tomato juice and encouraged and supported the addition of vitamin C to selected fruit juices, synthetic and imitation fruit beverages, and soft drinks as important low cost sources of vitamin C, as Canada has done with apple juice.

Today we are at the beginning of the era of imitation foods and meal replacements, and one of the immediate questions now is whether to add lysine or methionine to suitable foods while threonine and tryptophan are within the realm of possibility. We already see the same questions, doubts, and uncertainties arising that we saw years ago when the question of the advisability of adding synthetic vitamins to foods arose. Many of the same questions can be anticipated, such as the following:

What level of the amino acid should we add? Will the addition create an imbalance that will be harmful? Will there be toxic effects if too much is added? Should the mixture be adjusted to meet human needs or to an amount that would make a theoretically good protein? Some scientists will ask for experimental proof that the addition is effective before it is done. Others will say that it is too expensive and a waste of money, and that the same effect could be obtained by education or by making natural high quality protein available.

To my mind, many of these questions do not need an answer and to some it is impossible to reply. It has been amply demonstrated in experimental animals that if the nutrients are complete, except for lysine, the animal will not grow normally until lysine is added. It would appear that given a population on a diet of low protein quality in which lysine is one of the limiting amino acids, that nothing but benefit could be obtained from the addition of this non-toxic nutrient to a commonly eaten food containing protein of low biological value in which lysine is an important limiting amino acid. To try to set up a human experiment to prove this advantage is exceedingly difficult because of the difficulty of controlling the entire diet and the inability to maintain adequate controls.

Ethically, one cannot withold lysine from a food supply to human beings when animal experiments indicate that the results can only be deleterious. It is also evident that if a diet is deficient in several nutrients, as is the usual case in human diets, the provision of only one of these nutrients such as lysine is going to have little or no demonstrable effect. It would seem practically impossible to find a population existing on a diet deficient only in lysine.

It seems to me that one can only take the view that we must do everything we can to make our food supply nutritionally adequate at the lowest possible cost. In order to do this, we are going to have to use synthetic nutrients of all kinds on the basis of demonstrated safety and effectiveness in experimental animals. If there is a demonstrated need and it is economically feasible under suitable regulations, I can see no objection to improving the nutritive value of any food that is known to be nutritionally defective.

More synthetic nutrients are in the offing and there is growing evidence that they should be used. For example, considerable evidence can be cited for the possible advantage of adding synthetic vitamin B_6 and folacin to suitable foods. Other amino acids will quite likely become available and the chemical industry offers the possibility of further nutritionally improving the food supply as the world population continues to expand to the point where the distribution of natural foods of good nutritional quality becomes uneconomical as compared with synthetic products. One can see the possibility in the future of a way to make synthetic fats and possibly carbohydrates, as well as complete mixture of amino acids or mixtures of essential amino acids and non-specific nitrogen, to supply the calories and nutrients we may not be able to obtain

in any other way.

So far all efforts to change food habits by educating people to eat more nutritious products have failed generally when the more nutritious product has a changed appearance or taste, or violates some social or religious taboo. The addition of synthetic nutrients has been accepted now in the United States for more than one-fourth of a century. Its acceptability and effectiveness has been amply demonstrated in iodized salt, vitamin D milk, vitamin A fortified margarine, and in enriched bread and flour. Synthetic nutrients must be added to foods in a way which does not disturb the primary motivations to buy the food.

Although people buy food primarily on the basis of taste, appearance, habit, and cost, a new factor is appearing in the trend to reject processed or fortified food in favor of so-called natural foods. One of the causes of this appears to be an unwarranted fear created, at least in part, by government actions such as the withdrawal of tuna fish because of the mercury scare.

Food mixtures and meal substitutes of various kinds—which also may be nutritionally improved and made economically feasible by the use of synthetic nutrients—pose special problems. About the only practical way to make rapid and effective progress is by adding nutrients to our accepted, widely used, low-priced foodstuffs, or introducing nutritious imitation food products which simulate in organoleptic characteristics higher-priced and less available products with which we are familiar and which are widely accepted. Extensive and broad progress has been made in this direction during the past few years and the research laboratories are continuing to turn out an increasing number of promising products.

In conclusion, I would like to point out that people eat meals—not foods. In my opinion it is a mistake to try to make a "perfect" food by trying to make one product nutritionally complete in every respect. People have traditionally eaten meals of different foods and it can be anticipated that they will continue to do so. Our goal should be to develop foods of good nutritional value that can be combined into attractive meals which will meet the nutritional needs of the individual.

Our past experience with fortification makes it clear that the future in this area is most promising and essential for our health and well-being, and must be developed with careful planning, understanding, and regulations in order to make it both effective and safe.

Regulatory considerations in fortification of processed foods, as in other things, must rest on some sort of philosophic base if they are not to become a meaningless jumble. Accordingly, it is appropriate that we examine that base.

First of all, the Food and Drug Administration has long accepted the premise that it is quite feasible to maintain a satisfactory level of human nutrition from unmodified natural foods. This principle can hardly be questioned; the continued survival and well-being of the human race is fairly strong supporting testimony. It does not follow, however, that all people are now being, or have ever been, well nourished from their habitual consumption of natural foods. Within the memory of many living today there was widespread deficiency disease in this prosperous country, and recent surveys have shown that while deficiency disease, except for goiter, may now be almost entirely gone, there is still an undesirable incidence of identifiable malnutrition.

The second principle that must be accepted, therefore, is that just as it is possible to be well nourished by a self-selected diet from the American marketplace, it has also been demonstrated possible to be ill nourished by a self-selected diet from the American marketplace. Three obvious reasons for the existence of malnutrition are poverty, ignorance, and indifference. These are not mutually exclusive. Two of these conditions, or all three, can exist in the same consumers.

The national administration has committed itself (1971) to the objective of abolishing hunger and the associated malnutrition from this country. I have taken the liberty of modifying the commitment to include the malnutrition associated with hunger, and to exclude that which is so widespread in the form of obesity associated with overnutrition. Accordingly, any regulatory actions need to be taken in view of this commitment. The factor of poverty may be bypassed for the present discussion, not because it is not important, but because it is not germane. Actions directed at this problem are in progress, but are not the direct concern of the Food and Drug Administration.

The factor of ignorance is obviously an important one in generating malnutrition. Surveys have shown malnutrition among people who could afford an ade-

chapter 9
regulatory consideration in fortification of processed foods

by
Virgil O. Wodicka, PhD

quate diet. They have also shown good nutrition among people at low income levels. Poverty is obviously not the whole story. It may well be questioned, however, what concern regulatory agencies have with ignorance.

The concerns of the Food and Drug Administration lie largely in the area of labeling. As our population becomes increasingly urban, more and more of us are buying our food rather than growing it and increasingly we are buying it in packages. Statistics also show an increase in the proportion of this food that is processed and, therefore, usually in the type of package that does not permit viewing the food. Accordingly, it is necessary to communicate to the consumer the identity and characteristics of the product in the package.

In the past, much attention has been

given to a precise name for the product and to a painfully explicit and detailed list of ingredients. Little or no attention has been given to the nutritional properties of the product, however, unless they arose from added pure nutrients. Even here the major attention was given to the nutrients added and not to the total amount present. This brings us around to the second philosophical principle that is now guiding the regulatory efforts of the Food and Drug Administration. This is that the consumer is entitled to know the nutritive value of the product she purchases.

In days past, when even processed foods were sold as commodities, the communication of nutritive value on labels was less important. Those with a formal knowledge of nutrition knew the nutritive value of the commodities and could select their diets based on this knowledge. Those without a formal knowledge of nutrition had traditional food habits evolved over a period of centuries that gave some assurance of good nutrition even though the reasons for it were not understood.

In our melting-pot society, ethnic food habits have been badly eroded. In addition, commodities that may have been important in filling critical gaps of the nutrient pattern may be difficult to get in today's US economy. Furthermore, the trend toward convenience foods and away from commodities has increasingly put the control over nutrition in the hands of the processor rather than the purchaser. With highly formulated foods, neither the person trained in nutrition nor the follower of folk tradition can tell the nutritive value of the products.

The Food and Drug Administration has launched a program of nutritional labeling. This program will not be mandatory, and it is, in fact, doubtful whether the legal authority exists for it to be so. On the basis of discussions with key factors in the food industry, however, there is good reason to believe that the industry leaders will undertake nutritional labeling, and, in fact, many of the food chains are already impatient to get started. It now appears likely that so many companies will enter this program that few will dare to stay out.

In spite of the fact that the Food and Drug Administration started work on this program a year ago (1970), there has been no overt action in the form of regulations. This is because there has been a desire for some degree of assurance that the labeling

pattern offered would communicate and not just confuse. If the labeling program is to ameliorate the malnutrition of ignorance, it must be reasonably simple and understandable. It must communicate not only to the trained nutritionist or physician, but also to the housewife without much training in formal nutrition or even very strong motivation.

Several possible approaches have been developed and are about to be tested for their effectiveness in communication. They have already been reviewed by representatives of various segments of the industry for feasibility.

In spite of the fact that nutritional labeling—when the regulation is issued— will not be mandatory, the regulation will not be toothless. If a nutrient is declared on the label and the food in the package does not, in fact, supply the declared amount of nutrient, this constitutes misbranding and is eminently actionable under the law.

Accordingly, it will not be enough for the manufacturer to calculate what his food should contain of the various nutrients based on tabulated average values and his recipes. The label is intended to apply to the actual article of food offered for sale and not to some theoretical prototype. Some measure of quality control is obviously called for. The Food and Drug Administration is already making plans for surveillance and enforcement to follow on the heels of its labeling regulation.

There are other programs in government to deal with the factor of nutritional ignorance and, in fact, the labeling program will be of much diminished value without such efforts. This discussion, however, is limited to the regulatory phases of the overall attack on the problem. Accordingly, it is appropriate to pass on to consideration of the factor of indifference. If the consumer is not much interested in nutrition, efforts at education and communication will have little effect. This is where fortification steps into the spotlight.

The White House Conference on Food, Nutrition, and Health recommended that the Food and Drug Administration develop nutritional standards for foods. The legal authority for this probably exists under the provision for US standards of quality. On the other hand, action by such a mechanism would be extremely slow and cumbersome. A standard of quality would have to be developed for each food concerned and would be subject to all the usual administrative and procedural delays, including the possibility of hearings. To have any noticeable effect in the next five years, this operation would require a large increase in staff and budget.

Accordingly, this approach did not appear very promising. On the other hand, the concept embodied in this recommendation has obvious merit in dealing with the factor of indifference. If a food is required to supply stated levels of nutrition, then it is not critically important whether the consumer knows or cares that the nutrition is there. Such an approach is needed to cope with this third major cause of malnutrition.

The path chosen was the development of a set of nutritional guidelines for classes of foods rather than individual foods. Proceeding by class would reduce the number of sets of guidelines to a manageable number in place of the forbidding list of individual foods. At the same time, it would tend to provide a considerable degree of nutritional interchangeability within class and thereby make dietary selection less critical. This program, like the nutritional labeling program, would be voluntary rather than mandatory; but here again, conversations with leaders of the industry strongly suggest that there will be an aggressive movement to follow the guidelines when they are issued. Once again, it is to be expected that the forces of competition will take over, once the leaders have made the move.

The Food and Drug Administration established a contract with the National Academy of Sciences to obtain the recommendations of the Academy on what classes of foods should have guidelines and what the guideline values should be for the various nutrients. The Academy has established a special committee under the Food and Nutrition Board to study this question and make recommendations, and guidelines for the first two classes are close to publication. These are complete dinners and formulated main dishes.

Implementation of the guidelines again brings up philosophical questions. The guideline situation is not vastly different from that of enriched flour. In the case of enriched flour, there is actually a standard rather than a guideline and the standard is a mandatory one. On the other hand, the issuance of this standard did not cancel the standard for flour that is not enriched. In this sense, therefore, compliance is voluntary. A miller may offer to the consumer either enriched or unenriched flour. The fact that almost all flour sold to households is enriched is a hopeful indication that behavior under the guidelines will be similar.

The parallel with enriched flour, however, highlights the philosophic question. The standard for enriched flour was deliberately set with nutrient levels above those normal to whole grain. This action was taken as a measure of public policy to assure the widespread intake of the nutrients offered, with or without the conscious action of the consumer. This step is widely credited with the virtual elimination of clinical deficiency disease in this country. It may be reasonably questioned, therefore, whether the Food and Drug Administration has a similar intent with the guideline program.

As things now stand, this must be answered in the negative. The purpose of the guideline program is to provide to the consumer those nutrients that the consumer might have reason to expect in the food offered and at the levels that might be expected. The principle is one of guaranteed nutrition rather than one of nutritional supplementation or medication. The next principle of policy that should be set forth, therefore, is that in achieving guideline nutrient values, first preference should be given to reaching the specified levels with natural foods rather than by supplementation.

This is for the obvious reason that there is not yet a high degree of assurance that all essential nutrients are known, and in any event, there is a disturbingly long list of vitamins and minerals for which firm requirement figures are not yet at hand. The present state of nutritional knowledge will not yet support moving with confidence to a fully synthetic diet and any step in that direction must be taken with caution.

Although preference is given to the achievement of the stated nutrient values with natural foods, second choice is supplementation with specific nutrients to reach the stated levels. Emphasis will be given to nutritive value of the food as consumed with only secondary consideration accorded to the source of the nutrient. If only the listed nutrients were of concern, of course, the source would receive no attention at all.

One of the major developments stimulating the initiation of the guideline program is the rapid progress of technology in manipulating the properties of foods to permit the achievement of a particular set of performance characteristics from a wide variety of starting materials. This has been conspicuous in the demonstrated ability to match the properties of foods of animal origin with others consisting predominantly of plant material. The achievement of a close match in sensory properties, however, does not automatically imply a close match in nutritional properties.

There is the further prospect that the introduction of these formulated analogues only paves the way toward the widespread use of materials of plant origin to fill roles traditionally filled by foods of animal origin even though the resemblance between the end products dwindles year to year. In other words, a fairly close match must be achieved initially to bring about the substitution, but this match becomes progressively less necessary as the years go on.

Accordingly, matching nutritional properties between the analogue and the model is not a very satisfactory guide for a permanent regulatory stance. It becomes virtually mandatory to go back to basics and develop nutritional profiles based on the needs of the consumer rather than on traditional consumption patterns. This is likely to involve attention to amino acid patterns, as well as vitamins and minerals. It may also involve special attention to fat chemistry.

The fact that the regulatory focus is shifting from a base of established food habits to a base of food and nutritional science must not be taken as license to the hucksters to add unlimited quantities of pure nutrients to everything to give an added kick to an advertising campaign. It is necessary that the regulatory program guard against excesses as well as deficiencies. As the first step in this direction, a regulation has been drafted converting the amino acids from substances generally recognized as safe, to regulated additives. This will permit closer control of their use in nutritional enhancement. This regulation has now been published as a final order as §121.1002, Part 21, *Code of Federal Regulations,* on July 26, 1973.

The Food and Drug Administration will take whatever steps are necessary, including the seeking of additional legislation if it comes to that, to see to it that the addition of pure nutrients to foods is kept on a sound scientific basis and not used as a form of one-upmanship. The Federal Trade Commission is already beginning to exhibit its convictions along the same line. The Federal Trade Commission and the Food and Drug Administration have increased their efforts at coordination and, accordingly, it is certainly to be hoped, and probably to be expected, that the two agencies will maintain a reasonably consistent policy between the standards, labeling, and additive policies of the Food and Drug Administration and the advertising policies of the Fedeal Trade Commission.

To sum up, the regulatory prospect with regard to fortification, or more specifically, the addition of pure nutrients to foods, is that there will be considerably less restriction and, in fact, some encouragement of this practice where it is necessary or desirable to guarantee an appropriate level of nutrition in the food as eaten. The proper communication of the nutritive value of this food will be strongly encouraged. Adherence to the guarantee will be enforced. On the other hand, there will be continued, and perhaps even heightened, pressures to keep the forces of competition from generating products with inappropriately high nutrient levels. Every effort will be made to preserve the practices of traditional good nutrition while taking full advantage of today's science and technology.

Since 1920, we've greatly improved our capacity to produce food and we have met the demands of the population with attractive, safe, and convenient products. But the most important facts about food production and nutrition are not getting to the consumer. While most people seem highly sensitive to food problems, they do not have an equal awareness of the relative nutritional worth of today's foods. If professional nutritionists are communicating well with each other, and sometimes I doubt this, they're excluding physicians, dieticians, food buyers, and, certainly, meal planners, from their conversations.

What's lacking is effective communication of nutritional information to the consumer. This can be considered the most important problem confronting nutritional science today. What is needed is a clear, more precise way to define qualitatively what a food is in terms of human nutrient requirements.

The Food and Drug Administration may eventually forego its references to Minimum Daily Requirements (MDR). There is also reason to hope that the Food and Nutrition Board may begin to treat the Recommended Dietary Allowances (RDA) more flexibly. The Board may even extract a standardized dietary allowance, recognizing that, in fact, the meal planner and food buyer are concerned with groups of people. They're buying food for families, and while individual differences in nutrient requirements have academic interest to the nutritionist, they are of little practical worth to the consumer.

If the Food and Nutrition Board would standardize the Recommended Dietary Allowances, this could be the basis for adequate food descriptions that could be useful to a physician, to a dietician, and maybe even to a meal planner. Such a document, however, should recognize that the person planning meals probably knows little about grams (gm), milliliters (ml), micrograms (μg), milligrams (mg), or international units (IU).

I have attempted to produce an RDA that allows for age and sex differences in a practicable way. For every 1,000 calories, for example, there ought to be about 30 gm of protein, 2,500 units of vitamin A, 400 mg of thiamine, 650 mg of riboflavin, 25 to 35 mg of vitamin C, 400 mg of calcium. My number for iron (8 mg per 1,000 calories) has been greatly influenced by the data presented by Dr. Combs

chapter 10
caloric requirements in relation to micronutrient intake

by
R. Gaurth Hansen, PhD

(in Chapter 1) in respect to the National Nutrition Survey.

Calories can provide a common denominator for expressing food value. Nutrient requirements are not always strictly proportional to calorie needs but calories are the most likely common denominator for qualitative descriptions of various foods. It doesn't really matter in what form food is consumed, as long as total calories correlate with adequate quantities of all other nutrients.

Using a calorie-based system and data from the US Department of Agriculture food surveys, one can describe foods that are commonly available in the US. A consumer has an infinite variety of possible choices that can either balance or unbalance his diet. Generally available within that infinite variety of choices are foods that will meet an individual's nutrient requirements.

If it is assumed that all available food in the United States is ground in a Waring blender and then chemically analyzed for its caloric and other nutrient content, and a standardized RDA can be compared with the nutrient analysis, with calories as a common denominator, the result is a nutrient to calorie ratio or nutrient density. As illustrated in the chart (Fig 16) such a process indicated that the proportion of protein to calories is certainly adequate. The ratio of vitamins to calories is adequate, and, in all cases, when calorie needs are met we have 1.5 to 2 times more of all the nutrients than is necessary. The first questionable item is iron. Remember, for iron I've used a ratio of 8 mg per 1,000 calories. Hence for all individuals to receive an adequate intake of iron, particular care needs to be exercised to make the proper food choices.

This same nutrient-calorie ratio concept may be applied to the four individual food groups on which the educational program of the US Department of Agriculture is based. In the meat group (Fig 17), the ratio of protein to calories is almost three, while vitamin A is one, and thiamine, riboflavin, and iron are almost two. The deficiencies are in ascorbic acid and calcium.

According to this concept, the unique contribution of the breads and cereals (Fig 18) is thiamine. We also see the importance of enriched cereals as sources of iron. This presentation says nothing, however, about the quality of the protein provided.

The milk group (Fig 19), in addition to a two-fold excess of high quality protein, furnishes a good part of the riboflavin and calcium in the average American diet. You see the uniquely low concentration of ascorbic acid and the deficiency of iron.

Applying the same system to fruits and vegetables (Fig 20), their exceptional contributions to the diet of vitamin C and vitamin A are revealed. Once again, the system doesn't say anything about protein quality. Since the calorie contribution of fruits and vegetables tends to be low, this system tends to overstate their nutrient contributions to an average diet.

When a typical breakfast is analyzed using this system, it is obvious that an orange (Fig 21) makes an exceptional contribution of vitamin C, while providing a balanced supply of the other nutrients.

An unsupplemented breakfast cereal (Fig 22) has calories in excess of other nutrients, except for thiamine. Few people consume cereal alone; however, the nutrient composition of whole milk (Fig 23) fortunately has complimentary strengths. Combining milk and breakfast cereal (Fig 24) in their usual proportions produces a much more nutritionally attractive breakfast item.

Looking at nutrient-supplemented cereal (Fig 25), on the other hand, it seems there are almost no limits on what can be done. The processor merely has to decide what foods to supplement, and to what extent. I'm not sure that the nutrient-calorie ratio ought to be either far in excess of one, or very much less than one.

In the instant breakfast—a meal replacement—again there's no limit to what one can do with chemically synthesized nutrients. Again, the chart (Fig 26) indicates nothing about protein quality, and nothing as to what nutrients these products *do not* contain. This point becomes critical as more people consume more of their calories from snack foods (Fig 27). We have to make sure that the rest of our total diet provides the nutrients that foods of this kind do not incorporate.

We certainly also must be aware of what foods other than snacks *do not* contain—for example, trace minerals. Our nutritional planning must take into account these other elements. It's becoming clear that they are important in the diet. But we do not know at what point they may become limiting. While we may not see (or readily identify) frank deficiency symptoms, we certainly want to be sure that average diets prevent such phenomena.

In summary, then, it is critical that we develop a means of better communicating nutrition information. I've suggested, I hope with offense to nobody, that the Minimum Daily Requirement may have outlived its usefulness as a nutritional statement. I've also attempted to present the Recommended Dietary Allowance in a form that should be helpful in describing food quality in terms of human needs for nutrients. From these two parameters, the nutrient density or nutrient to calorie ratio emerges as an index of food quality. Data are available that could provide some tentative basis for extension of the Recommended Dietary Allowance to include other nutrients. I have also emphasized that the increasing popularity of snack foods makes it imperative that we consider what nutrients various foods do *not* contain.

Finally, we must convince consumers that eating a large variety of foods is a sound practice.

Figure 16.
Aggregate Food Available.

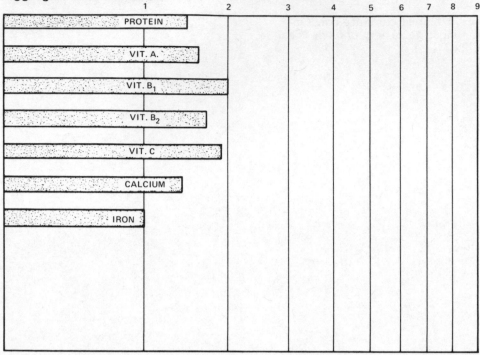

Figure 17.
Meat Group.

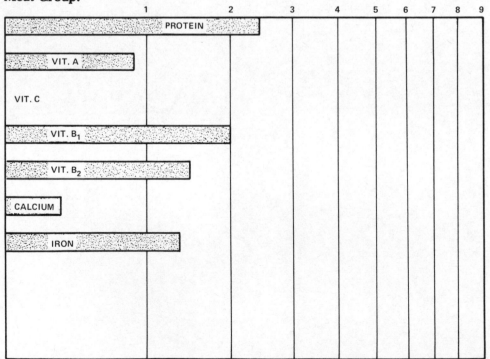

Figure 18.
Bread and Cereals.

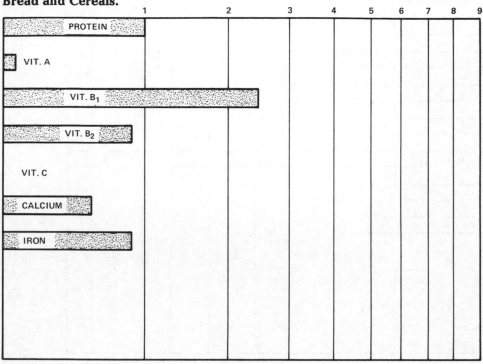

Figure 19.
Whole Milk.

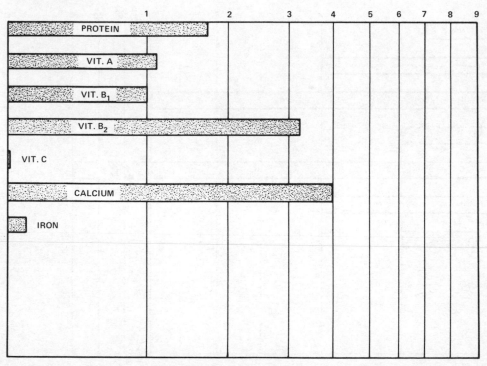

Figure 20.
Fruits and Vegetables.

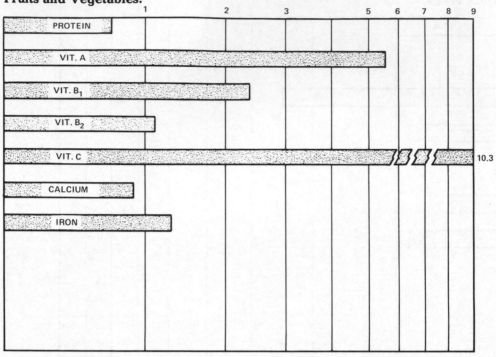

Figure 21.
Orange.

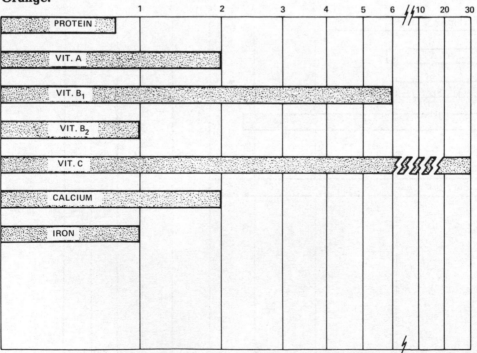

Figure 22.
Unsupplemented Cereal.

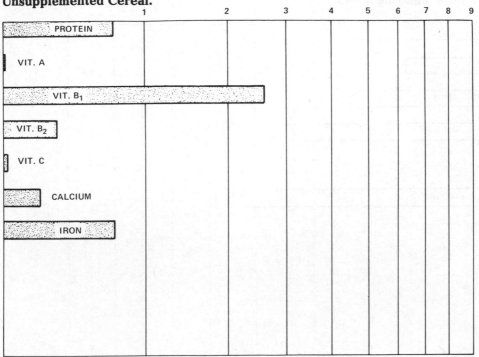

Figure 23.
Milk Group.

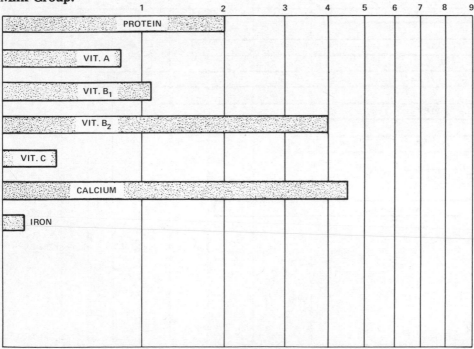

115

Figure 24.
Cereal Plus Milk (Equivalent Basis).

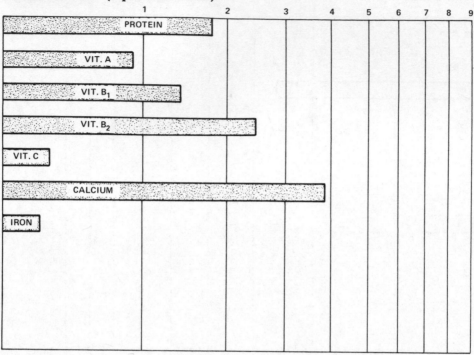

Figure 25.
Nutrient Supplemented Cereal.

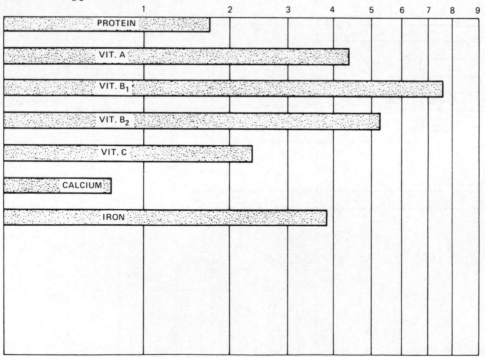

Figure 26.
Instant Breakfast.

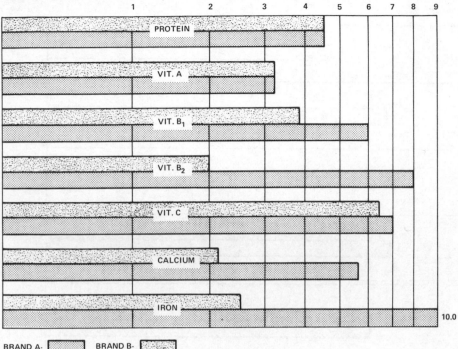

BRAND A·   BRAND B·

Figure 27.
Breads, Flours, and Similar Products—Per 100 Kcal.

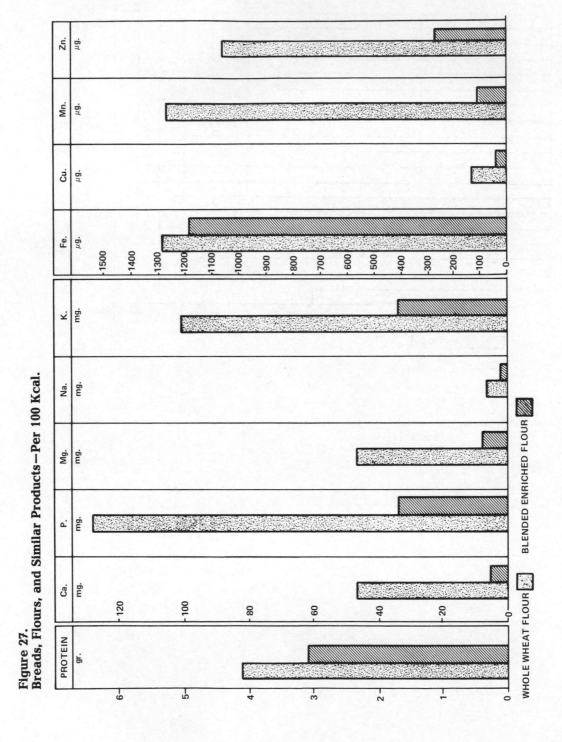

WHOLE WHEAT FLOUR

BLENDED ENRICHED FLOUR

118

More than a generation ago I first learned that the history of nutrition could be described as changing a singular noun (nutrient) to a plural noun (nutrients). I believe it was approximately 1840 that the English chemist, Prout, found that foods could be separated into three chemical components, and suddenly there was fat, protein, carbohydrate, whereas previously there was only nutrient in foods. Minerals as essential components of nutrition were discovered about the time of our Civil War. Vitamins came along at the turn of the century. Today we know of some 50 nutrients, and no single food—not even mother's milk—contains them all in adequate amounts, and that is the physiological reason why we need a variety of foods in the diet.

But there is also a psychological reason for a variety of foods in our diet. Eating is one of the pleasures of life. Different flavors, textures, colors, and aromas are enjoyed. The same few foods every day make life pretty dull. Most of the "odd ball" "revolutionary" reducing diets of the eat-all-you-want variety are very limited in their choice of foods. For example, the infamous Mayo Diet (never any relation to the Mayo Clinic) of grapefruit, eggs, and bacon—nothing else, but eat all you want—does result in weight loss. This limited fare becomes so monotonous after 2 to 3 days that you have had it, you can't eat very much.

I believe it was during the First World War that nutrition educators dreamed up the Basic Eleven Food Groups—one or more of each group to get variety into the diet and hence all of the many nutrients needed. But too few people could remember 11 food groups, so during the Second World War this was simplified to the Basic Seven. Again, not enough of us could remember seven food groups, so we now have the Basic Four Food Groups. Other countries have basic food groups of varying number, built around their local foods and food habits, but all designed to teach variety in food consumption.

Despite the best efforts of nutrition education, we haven't made too many inroads in changing food habits, in getting people to eat a wide variety of foods from the various food groups. And if we have had little success in getting variety into US diets, with the great variety of foods available to those who can afford them, how will one ever get the masses of mankind in the

chapter 11
importance of variety in the diet

by
Fredrick J. Stare, MD

underdeveloped areas of the world to get variety into their diets, particularly when not much variety is even available?

With an expanding population in this country and in the world, and with a desire for less work in the kitchen, I think the way we will meet and overcome most of these nutritional challenges is through an expansion of the intelligent and continual safe use of preservatives, food additives, and the fortification of foods, particularly the three cereal grains that nourish most of mankind. These are rice, wheat, and corn, and they can be fortified with appropriate vitamins, minerals, and amino acids. They can be supplemented with various legumes or fish protein concentrates. The rationale for fortification of foods is twofold:

1. Too many people do not consume a sufficiently and properly varied selection of foods to provide them with the desired

amounts of the some 50 known nutrients.

2. To prepare, preserve, and store foods from harvest until they are consumed, and to build convenience into many of these foods, requires a certain amount of processing and refining with some loss of nutrient content.

Both of these situations provide the rationale, and indeed the necessity, for fortification which has been an important practical development for improved nutrition, and which I think will be more important in the years to come. The classic examples of fortification are: iodine to salt, vitamin D to milk, vitamin A to margarine, three B-vitamins and iron to white flour used for bread, and fluoride to water.

Enrichment and fortification of a few basic staples helps give assurance that good nutrition is readily available to all via every grocery store or supermarket. So-called "health food stores" with their "health foods," "nature foods," and "food supplements" are not necessary and, in my opinion, are a great waste of money. The claims that are frequently made for many of these products are deceptive, fraudulent, and represent quackery and charlatanism at their worst. We certainly are not going to nourish an expanding population on "natural foods" of a hundred years ago.

We must become more aware of the long-range effects of food and its nutrients on man. More people in the developed areas of the world can today get better food on their dinner tables because of technological breakthroughs in food processing. These nutritional advances must be brought to many more people. We may have to change some rather basic assumptions about what is sound nutrition. We all know, for instance, that we can increase the growth of the human infant by overfeeding him. But this is not necessarily good; in fact, it is probably bad from a health viewpoint.

Since most of mankind lives off of cereals—primarily rice, wheat, and corn— the production of more of these foods is important in the near future. The "green revolution," meaning mainly higher yielding strains of the basic cereals, is a big step forward. But the protein of these cereals is low in one or more essential amino acids— rice in lysine and threonine, wheat in lysine, and corn in lysine and tryptophan.

Again, genetic research may in time develop new strains not only with greater yield but improved protein quantity and quality. But today, not tomorrow, with the help of the chemical factory, it is possible to make a bread, or any other wheat product, with protein of good nutritional quality by the addition of the single missing amino acid, lysine. Wheat products made with lysine do have improved protein quality and are economically feasible. Taste and cooking qualities are not affected. Food habits do not have to be changed.

Thus, I feel quite strongly that man's nutrition in his new life environment—with far more men—will depend more and more on fortified foods, particularly fortified cereal products. I might add fortified potatoes and fortified sugar, cheap and efficient sources of energy— and energy is a basic need, perhaps after water, the most basic requirement of our diet. We will be fortifying these basic foods, possibly even tea, coffee, and soft drinks, not only with vitamins and minerals, but also with protein concentrates, amino acids, and even certain fatty acids.

All of this fortification of foods involves the addition of chemicals to foods. All consumers are made up of chemicals, even those who object to the addition of chemicals to foods. Some of the chemicals added to foods are not nutrients but are added to preserve the nutrients of the food, or as in the case of monosodium glutmate, modify the taste, and hence acceptability and enjoyment. Please remember that eating is one of the pleasures of life as well as a necessity.

In current parlance, food can be said to be a "gut" issue. Many persons, in the concern over pollution in the conservation movement, in the "hippie culture," and some in their concern for personal publicity and ego building, are yearning to return to a more "natural" existence, particularly in regard to food. This, of course, cannot be. By the end of this century, the population of the world will exceed 6 billion. The matter of population control, therefore, is an extremely urgent one. But even success in this area does not permit traditional or "natural" food sources and traditional methods of agriculture, food production and processing, and distribution to avoid mass starvation. The use of unconventional food sources, the use of food additives, and the use of new methods of preservation, packaging, and distribution will continue to grow as a matter of our sheer survival. Along with it must grow, of course, our concern, constructively expressed, for food safety.

Consider the proper question to be the safety and nutritional value of the food in the total diet and the existence of the individual, rather than attempt to judge ingredients by themselves as safe, unsafe, or nutritious. Too many times we see references to the desirability of "thorough testing to prove that an ingredient is safe." But this is at least a triple fallacy.

In the first place, the testing is almost always carried out in animals, in which the response may be different and sometimes drastically different from the response in man. However, such testing is a first step in evaluating safety. Secondly, one can never prove safety. One can only conclude that in a given experiment, no evidence of harm was detected. This does not mean that in a slightly different experiment, perhaps carried on for a somewhat longer period, or with a different investigator, some evidence of harm might not have been found. Finally, every component in our diet, and even "pure" air, is capable of causing us harm if taken in excess or under the wrong circumstances. We require salt, but everyone knows of people who have died from drinking sea water because of its salt content. Cobalt is an essential mineral nutrient in our diet; indeed, it is a part of vitamin B_{12}, but it provides one of the very few instances of a food additive which was implicated in the death or injury of humans who consumed it. It is worth noting in this case, however, that these people did not consume it in normal amounts. This cobalt was added to beer as a foam stabilizer at a level of about 1 ppm, and the injury or death was restricted to those people who were unwise enough to drink from 6 to 12 quarts—not glasses of beer per day.

These similar examples simply illustrate *that there really are no safe or unsafe substances. There are merely safe and unsafe ways of using them.* We must really consider the safety of food, not of substances added to food or contaminants from the environment. We tend to lose sight of this in most discussions.

In man's new environment we will depend more on foods preserved, flavored, and fortified with added chemicals, many of which are nutrients but many of which are not. Not only are many essential nutrients capable of causing harm in the acute sense, but they may also do so in a long-term or chronic sense. We have overwhelming evidence of the hazards of overnutrition to the extent that the intake of too many calories and too much saturated fat and cholesterol has an effect on the incidence of cardiovascular disease.

This information comes to us not from testing on animals where the response may or may not be relevant, but from experience with the organism of most concern to us, the human being. But the most important aspect of diet and cardiovascular disease is that one can manipulate the diet with some ease. Many of the "preventive principles" of nutrition can actually be built into manufactured foods, and most of the foods in the developed areas of the world today, and more in the future, will be manufactured foods.

Sophisticated consumers of the future will not only feed themselves better; they will be far less prone to be the victims of ignorance, the misplaced enthusiasm of the food faddist, the hucksterism of the charlatan, and the malice of those who inflate their egos and earn their living by alarming others. There is much to be concerned about in the years ahead and our efforts should be expended on the right points. And one of these right points will be an orderly and intelligent expansion of the concept of food fortification. The rationale for this is too many of us— and far more in other parts of the world— do not consume, or have the opportunity to consume, a variety of foods that provide in adequate amounts the some 50 nutrients known to be necessary for good nutrition.

In summary, a proper variety of foods, specifically, a variety selected from the various Basic Four Food Groups will provide the best nutrition each of us can obtain genetically. Intelligent fortification of certain key foods will assist in obtaining this adequate nutrition by those who do not or cannot consume this desired and proper variety of foods.

Rapidly changing food patterns in the United States have altered our thinking as to what we can or cannot do in nutrition. Nutrition experts a few years ago would have said that fluid milk should be looked upon as a product that provided a certain percentage of calories and certain nutrients. Bread formed a significant portion of the diet of many people, and as a class, the cereal products could be expected, in the diets of most Americans, to provide certain food components.

When these statements could be said with some authority there was relatively little consideration given to so-called "snack foods" or institutionally prepared foods, or the meals that were consumed away from home. Meat, vegetables, potatoes, bread—all easily identified by the consumer—could be looked upon as forming the basis of our meals, and meals were consumed at home.

Today, fabricated main meal items, complete meals, liquid and solid meal replacements, meat analogs, and a multitude of aerosol cans containing food products have replaced more traditional foods. They make it extremely difficult for the consumer and, in many cases, for the professional to clearly identify how to place foods in terms of food classes.

Many processors try to associate new food products with a traditional class by their appearance, but not necessarily by their nutritional quality. There is also the problem, in terms of class identification, of foods consumed outside the home. Some are high calorie meals, high in fat, luxury meals. But a meal may also be a hamburger and french fries—a combination which has become, in fact, important for many Americans. To be sure, many of these meals are nutritionally sound and make a very important and significant contribution to the diet of those who are consuming them. But, I think we'll all agree that for many individuals, the hit-and-miss pattern of too many calories and not enough foods containing a variety of nutrients can lead to potential problems.

And then, snacks. I don't know how to define a snack. In our house, it's anything the kids can sneak out of the refrigerator when my wife and I aren't looking. It may be a very good snack, a piece of cheese, carrot sticks, perhaps a glass of milk and some left-over chicken from yesterday. Perhaps it will be some crackers or potato chips or candy. The

chapter 12
classes of foods which could be considered in defining nutrient intake and needs

by Ogden C. Johnson, PhD

word snack has been misused to a point that none of us know exactly what it means.

These snacks, or little meals, are making an increasing contribution to the caloric intake of many Americans. As we can't define them, we can only say that they may contribute little or much to the total nutrient intake. As snack meals become more important, we must give consideration to what their nutrient content should be, and whether they should be put into individual classes or treated as if they were all the same.

Defining a food class or food group is very difficult. In the past, 11, 7 or 4 food groups have been very common as planning denominators. They were developed so that foods with similar appearance or similar use could be grouped together. We ignored the fact that the overlap was considerable in some cases and that there were always foods in this group or that group which simply did not make the contribution expected of the foods in the group. Food groups proved useful in setting up nutritional guidelines for nutrition education programs. And, as long as our eating pattern followed those kinds of food groups, I think we had no major problems.

In recent years, the classical groupings of foods have become blurred. It is more and more difficult to put formulated products in a class. In some cases, a formulated product represents three or four classes. In some instances, you can't identify a traditional class. When the individual consumer no longer can place foods into appropriate classes, he begins to do some classification on his own and he probably puts some foods into classes which are erroneous and consumes them on the assumption they provide certain things to his total diet that simply are not there.

There is another way of classifying foods that we all use. We group foods in association with meals. We have breakfast items and items for lunch and others that we think of as dinner items. But we all know individuals who quite routinely eat foods for dinner that you and I might consider foods for lunch or breakfast. Thus, as a classifying system, this leads to some basic problems.

We also may classify foods by associating them specifically with nutrients. This corresponds with the four-food or seven-food group classification but goes one step further and eliminates the identity of the food. A protein source can be meat or milk or a cereal product. It can be a fabricated food which bears little resemblance to any of these, or it can be a meal substitute in a solid or liquid form. A given food can fall into many classifications under this format, some of which are "natural," and some which have added nutrients.

As the concern for nutrition increases, there is a tendency for all of us to assume that people will seek general nutritional information. Perhaps we're wrong. Perhaps over the next 10 years we will, in fact, find that actual nutrient classification is something that consumers desire. An individual will have to be trained, of course, and there will have to be appropriate nutritional labeling. It is

probable that this approach will be successful, given time and an appropriate educational effort to orient consumers to nutrient content as a major type of classification.

Regarding the growing development of food analogs, I think the time will come when direct imitating of traditional foods will be ignored and the manufacturer will put out new products in forms and flavors to succeed on their own merits. Nutritional quality may be determined not by what you or I might consider to be the best, but how the manufacturer assumes the consumer will use the product or how his advertising directives say it should be used.

In fabricating food for which there is no traditional comparison, the temptation to make it better—to add higher and higher levels of nutrients—will be overwhelming. Yet, as many experts have stated, we need to resist this, to avoid toxicity and confusion.

Traditional classification of foods will become almost impossible to achieve as completely nontraditional foods become plentiful. We will have products that can be used as major sources of protein but carrying with them the nutrients expected in fruits and vegetables, and consumed, perhaps, as a fluid which resembles the milk group. Since the producer of fabricated items will not be bound by older concepts, we will have to consider the functions of the foods, and perhaps, open a door to a new type of classification.

We'll have to be aware of the possible requirements for those nutrients that we know very little about today. We may have to add nutrients to products before we are able to detail the nutritional requirements. It is extremely important that we add certain of the minerals, for example, before we have the absolute quantity that's needed.

There are many nutritional areas about which we know relatively little and have relatively little ongoing activity to generate new information. Detailed information on food consumption, for example, is not currently available. If we are to make meaningful decisions as to how to classify and fortify food, we must have information on how the public is consuming food. We must know what changes are taking place in food habits on a continuing basis. As new foods come into the marketplace, we will need a mechanism for determining how these foods are being used and what they are replacing. Are they replacing foods with like nutritive quality? How do we measure usage? Are methods used in the past acceptable? Can we measure both food intake and frequency of use? Are the methods that were developed for fairly simple products applicable to the complex formulated food mixtures in the marketplace today?

When we have food intake data other questions arise. Do we have good information on nutrient content? We also need to know what is the nutrient content of the traditional food products that are being replaced. Can we arbitrarily use values collected 20 years ago?

There was a time when nutrient analysis was attractive to the scientific community. I suspect this has not been true in recent years. The graduate students and their professors no longer look upon this as a useful activity; far more exotic and, therefore, more important things need to be considered.

Industry can be helpful through cooperative efforts in study of product use and composition. We need better information on the nutritional quality of food. We have been looking at nutritional information that is available. The industry can be and should be involved in providing new information. Many companies have nutrition information, but may be reluctant to make it available because they are unsure how it can be best utilized and fear it will be misused by the press or others. A cooperative effort can overcome some of these fears.

If we are going to talk about how one fortifies food and how one classifies food, we need to know what kind of nutritional problems exist. We need to know the nutritional status of our population on a continuing basis. There needs to be, in my opinion, consideration given to not just biochemical analyses or dietary recall data; we need to consider the performance capability of the population.

We cannot break out all of the performance factors that lead to increased infant mortality in low-income populations, but we must consider that nutrition could be one of them. We have not developed methods for doing much performance measuring in the broad sense. As we do, we may find that the so-called classical levels of nutrients do not meet certain performance needs, and that individuals vary more than we have been willing to accept. We also may find our requirement data is accurate, but the nutrients aren't being consumed.

We cannot force everyone to eat the same way. We cannot force all the foods into fixed groups. What we should be concerned about is the total nutrient intake, meeting needs regardless of the patterns that develop in eating. Over the next 10—or even 20 years—nutrition should come of age, and the role of nutrition—and consumer understanding of this role—be made a routine part of food practices.

In order to avoid the impression that remarks in response to the title question are principally a matter of personal opinion, it is well to outline the guidelines available to us for arriving at a conclusion as to which foods should be available with nutrient additions that make those foods more useful. The classical guideline, of course, is the statement of policy with respect to food enrichment that has been formulated and published jointly by the Food and Nutrition Board and the AMA Council on Foods and Nutrition. This not only lays down the guidelines for determining the suitability of fortification or enrichment, but states an endorsement of specific food categories that will meet these guidelines. This, as you are all aware, includes the cereals, fruits and fruit juices, and the addition of vitamin D to certain milk products, of vitamin A to margarine, and iodine to table salt. It includes a reference to the protective action of fluoride against dental caries and endorsement of the addition of fluoride to water supplies.

One of the important requirements in guides set forth in the statement of policy is that "the intake of the nutrient is below the desirable level in the diets of a significant number of people," and the second that "the food used to supply the nutrient is likely to be consumed in quantities that will make a significant contribution to the diet of the population in need." It is clear that the emphasis here is upon the assurance that there will be significant benefit to the consumer of the product in question.

This policy in the past has been followed closely and has been particularly helpful in the promulgation of food standards by the Food and Drug Administration. With respect to enriched cereal foods there are 13 specific items that have been standardized under the Food, Drug, and Cosmetic Act. Other standards include those for evaporated milk, nonfat dry milk with vitamins A and D, a series of canned fruit and fruit juices that may contain added ascorbic acid, and finally, there is the standard for oleomargarine with added vitamins A and D. It is not feasible to enumerate the foods with added nutrients now marketed for which there are no FDA standards. These, of course, are many and varied.

The recognized need to take into consideration the benefit to the consumer

chapter 13
what foods should be fortified?

by
Oral Lee Kline, PhD

encourages us to look at the categories of consumers that may be at risk from the standpoint of age groups. Experience not only in this country but in many underdeveloped countries clearly points to the segments of the population that may be at risk, those falling in these age groups—infants, preschool children, teenagers, middle age from the standpoint of obesity, pregnant and lactating women, and older age groups. If we look from another direction at populations at risk, as was done in the National Nutrition Survey, we find that the needs are greatest for those segments that are in poverty, many of those in rural areas, the migrant workers, Indians receiving government aid, and possibly the elderly living in apartments.

From both bases, it is reasonable and logical to identify the foods most appropriate for nutrient supplementation and most widely used within these categories and by the particular segments of the population listed above. To do so would meet the guidelines of the policy statement, providing the foods identified were suitable carriers and actually reach and are used by the groups of concern.

Some insight is also provided by the report of the International Working Group sponsored by the International Union of Nutritional Sciences (IUNS), the International Union of Food Science and Technology (IUFST), and the National Academy of Sciences (NAS). This conference, convened in August 1970, represented contributions from a number of countries and was organized by the Committee on Food Standards of Commission I of the IUNS. Here attention was given to the kinds of foods that are most likely to be of use in meeting the needs of groups at risk that we have enumerated.

The Committee on Food Standards stated in part, "It is recognized that in certain areas of the world food habits are changing rapidly and that a proportion of these populations no longer have three meals at home in the traditional manner" and "In other areas where most of the population still depends on traditional foods, we may anticipate the introduction of new food products which are necessary to improve the nutritional status of the people." Summary recommendations point specifically to the kinds of foods that are needed for improving the nutritional status of populations at risk and for which nutrient content must be assured. In four specific cases it was recommended that:

1. Enrichment of conventional foods be continued according to present policy to overcome specific nutritional deficiencies.

2. Nutrient content of fabricated (formulated) foods be related to their place in the diet.

3. Nutrient content of foods that substitute for or simulate traditional foods be similar to that of the food they are intended to replace.

4. Fabricated (formulated) foods used as meal replacements contain all the necessary nutrients in proportion to their caloric content.

The committee went on to state, "Increasing numbers of people, including vulnerable groups, are consuming snacks and other new foods at intervals during the day," and further, "There is need to examine the impact of these changes in dietary habits to develop nutritional standards for these new and formulated foods." Here we have an extension— beyond the list of standardized enriched foods already cited—that needs evaluation in view of changing food habits not only in developed but in developing countries.

This is a key thought to the problem that we face—not only innovation in food processing and food formulation, but innovation in our thinking of ways to assure fully balanced nutrient intake among all groups who are receiving adequate caloric intake. The problems of inadequate food consumption cannot be met by the enrichment of specific foods alone.

Another source of illumination is the recommendation of the White House Conference on Food, Nutrition, and Health. The panel on new foods had much to say about food fortification and in its first recommendation proposed an immediate food fortification program to provide to the public nutritionally complete foods to relieve undernutrition and malnutrition. The panel on food quality recommended that mandatory fortification either to the original nutrient content or above, where appropriate, be established for certain basic foods. The panel on food manufacturing and processing had much to say about enrichment and specifically proposed that the food industry accelerate its efforts to make available nutritious snack foods. A task force group under the heading of consumer protection would require that new foods should be equal in nutritional value to their traditional counterparts and priced coincident with their cost.

These are examples of the attitude of the panelists; some of these examples were answered by the agencies to whom they were directed. For example, the response to the recommendation of the panel on traditional foods reads, "There must be a continued effort in all government nutrition programs, including the school lunch program, to encourage the provision of milk and dairy products to children of all ages, and for pregnant and lactating women. Only instant nonfat dry milk fortified with vitamins A and D should be made available under the food distribution programs."

The proposal was accepted by USDA and they commented that their purchases of nonfat dry milk for donation to schools, day care centers, and low income families is nonfat dry milk that is fortified with vitamins A and D. The panel further recommended that where a new food purports to be a substitute for traditional food it should be required to provide equivalent nutritive qualities and also that a study should be made of the extent to which traditional foods are being modified and substitute foods are being formulated. The Department of Health, Education, and Welfare (DHEW) replied by saying that these recommendations will be implemented by the development of nutritional guidelines for food classes through collaboration with the National Research Council and industry groups.

A consumer task force group in getting to specifics with regard to the New Foods Panel's recommendation for a program of immediate fortification listed six basic foods that are to be fortified: bread, protein fortified teething biscuits, flour and corn meal, rice, two fortified processed meat products, and citrus and soft drinks. In responding to this recommendation, DHEW pointed out that nutritional guidelines, that will establish a national policy on fortification, are being developed by the Food and Nutrition Board.

With this background we can come to some conclusion with respect to foods that should be fortified. It is clear that such a listing would be in addition to standardized enriched foods that are now available. The question of whether or not there should or can be mandatory enrichment at all regulatory levels needs further consideration and cannot be answered here. Such an action, however, would have an influence upon the listing of other nonstandardized foods that would need to be enriched in the interest of consumers.

Following the needs of different age groups who may be at risk it is certainly important that for infants and preschool children milk products contain added vitamin D and nonfat dry milk products, both vitamins A and D. This applies also to the use of these products by women during pregnancy and lactation.

For the preschool child emphasis must be placed upon the provision of adequate protein which can be met by the use of cereal products appropriately supplemented with animal and vegetable protein. In this country great improvement would result, I believe, with the establishment of a standard for breakfast cereals that would require for fortified products the addition of nutrients that fall within nutritional ranges and that include protein supplementation to provide a significant part of the day's requirement of protein. This is an important step in view of breakfast food consumption patterns and in view of the wide acceptance and use of a variety of breakfast cereals. It is presumed that the Food and Nutrition Board Committee on Nutritional Guidelines would identify the appropriate nutrient limitations for this class of products.

For the teenager where caloric intake is sufficient, we have such a variety of food intake patterns that seem comparable only with respect to the use of snack foods. Rather than attempt a change in food habits in this age group it seems desirable to make certain that the so-called snack foods are made a significant source of nutrients that will supplement the amounts that would be supplied by the displaced traditional food. This nutrient list would include protein and, presumably, the enrichment formula used for standardized cereals.

Recognition of the fact that snack foods are frequently high in fat, and therefore the source of calories low in nutrients, should encourage consideration of a level of nutrients that is at least proportional to the caloric contribution. If we include with snack items the "between meal" foods or the traditional sandwich lunch combinations, we need to look at the nutrient contribution of these items and make certain that the calorie contribution from this source carries at least its proportional nutrient content.

With respect to the middle-age category, it is not likely that with the standardized foods now enriched and with those nonstandardized items that are fortified and available in the market there is a special nutrient need for this age group. A more important consideration is the provision of foods properly designated that are designed to assist individuals in reducing or maintaining an appropriate body weight.

Overnutrition is a common occurrence, with only a small percent of those overweight physiologically unable to arrive at and maintain the body weight conducive to good health. Concomitant with the need for foods of lower caloric density is the beneficial effect of supplements that provide bulk to the diet. We hear from time to time of attacks upon dietary patterns in developed countries that result in a predominance of "refined carbohydrates." This implies, of course, the reduction of the constituents of cereal grains eliminated in the milling process—which when consumed provide bulk.

Milling also reduces intake of certain dietary trace minerals that may well be of a significance not now fully identified. In the development of new foods and of food supplements it is surely possible for the food technologist to provide acceptable and safe food components that would make up these deficiencies.

For the old-age category where special foods are prepared that are particularly suitable for people in this category, there should be every effort made to be certain that the nutrient content of the modified foods is adequate for maintenance of good health. In consideration of the segments of the population at risk, either because of an inadequate caloric intake or because of necessity for reliance upon lowest cost foods, it seems clear that with the proposed food modifications there is no need for special foods or special enrichment with fortification formulas to meet the needs of people in poverty. This is a problem of delivery of food and of improvement in our methods of educating such people with respect to the appropriate food combinations that will provide optimum nutrition.

In summary, in addition to the enriched foods now the subject of FDA standards—enriched cereal products, fruits and fruit juices with added ascorbic acid, milk products with added vitamins A and D, and margarine with added vitamin A—the following are to be considered as of benefit to consumers. It is recommended:

1. That fortified breakfast cereals be brought into rational nutrient compositions according to an appropriate guideline, or preferably through FDA standardization with the requirement that added nutrients be of a kind and within ranges that have nutritional significance.

2. That there be continued emphasis upon the additions of vitamin D to fluid milk and vitamins A and D to dry milk products.

3. That snack foods, where feasible, be improved nutritionally by application of the enrichment formula and by supplementation with protein in amounts that at least are proportional to the caloric contribution.

4. That other "between meal" foods commonly used by teenagers be fortified to provide significant contributions of nutrients—for example, peanut butter supplemented with levels of vitamin A and protein to provide nutritionally important levels of these nutrients.

5. That attention be given to the problem of the effect of refinement of foods upon the reduction of diet bulk and that food technologists give consideration to means of effectively restoring this component to the diet.

6. That new foods, formulated foods and milk replacements, be supplemented to contain at least the nutrient content of the foods and meals they replace in the diet.

7. That further consideration be given to the requirement for the addition of iodine to table salt.

As a quantitative guide in carrying out these proposed changes and also in making useful additions to basic foods not now fortified, there needs to be a careful examination of nutrient losses through processing, an examination that extends beyond the present enrichment nutrients into the area of trace minerals, vitamin B_6, and protein quality. Such data will give us an opportunity to be more specific in upgrading the nutritional values of our foods in the marketplace.

The setting for education in nutrition varies from the formal classroom to a food editor's column with an aptly turned, easily remembered phrase. These extremes likewise illustrate the state of the art of teaching nutrition. It is proceeding somewhere at all times. Not all people are aware that they are being exposed to educational materials and subsequently pay little heed to it. Education in nutrition is handicapped by "counter educators" with pet concepts of nutrition or dietary supplements to sell. This places great burden on the legitimate educator who feels obligated to take time to set the record straight before getting on with positive education.

The basic concepts in nutrition education do not vary as greatly as do the settings in which it takes place. Nutrition education is generally based upon the tenet that meal patterns representing a wide variety of foods will assure adequate nutrient intake measured against a goal of performance such as the Recommended Dietary Allowance (RDA). Education is involved with teaching people how to select, prepare, and then to consume an appropriate variety of foods to realize that goal.

To make this concept work, certain staple foods have been enriched or fortified as determined by the needs of the population; other processed foods have nutrient levels returned by restoration. These classical methods of assuring adequate nutrient intake by the population-at-large were acceptable when the food supply was not so sophisticated. This was the case, relatively speaking, 30 years ago.

The theoretical result of controlled food enrichment is a food supply that can provide adequate nourishment for all people. The consumer has the responsibility of selecting and preparing food to take advantage of the opportunity. The educator has the responsibility to instruct the consumer in the selection and combination of foods into nourishing meals. Such educational efforts must, of course, take into consideration certain determinants of food choice: poverty, cultural and ethnic variants, prophylactic or therapeutic dietary situations, and the age of the consumers.

One popular teaching aid in nutrition education utilizes the concept of four basic food groups. The foods are grouped more or less by their common nutrient

chapter 14
education of the consumer

Philip L. White, ScD

characteristics (dairy products, meats and appropriate alternates, fruits and vegetables, and enriched and whole grain cereal products). The approach includes recommendations for the necessary number of servings from each group of foods to assure an adequate nutrient intake. This concept probably has its best utility in a person-to-person teaching situation where the opportunity is present to illustrate how combined dishes may contain foods from several groups. The sausage pizza is a classic example of a popular food item containing foods from all four groups— meat, cheese, tomatoes, and cereal. Yet pizza has also been used to illustrate the shortcomings of the food grouping system. Such grouping of foods is, in reality, a short-hand version of learning to combine foods without the need for detailed knowledge of nutrient composition of foods or of human nutrient needs.

135

TECHNIQUES OF CONSUMER EDUCATION

For purposes of orientation, the principal means of consumer education will be listed with no attempt being made to evaluate their effectiveness:

1. The media of mass communications: This includes radio, television, newspapers, magazines, and books. The material varies from 15-second public service announcements to scholarly books. It is in this area that the "anti-educators" find the fertile ground.

2. Formal school education: Formal education embraces teaching at all levels from kindergarten to postgraduate education. The quantity and quality of nutrition teaching is quite erratic.

3. Consumer outreach programs: This category is meant to include such activities as the USDA extension services and the Department's Nutrition Aide program. Included also would be public health and education programs of a person-to-person nature or those directed to small groups of people.

4. Education through professional services: This includes patient counselling by physicians, dietitians, and nurses.

5. Advertising: Advertising may be product oriented or may be of an institutional or educational nature. The latter is usually carried out by food trade associations. There is little doubt that advertising is the most influential force in consumer education.

6. Labels, package panels, and inserts: Labels can be vehicles for meaningful nutrition education above and beyond the information conventionally carried. Recently the prepared breakfast food industry has begun to publish educational messages on the panels of the package.

7. The back yard fence.

CRITIQUE OF CONCEPTS

Nutrition education falls short of the mark when it fails to conform to a logical flow of information and is devoid of the elements of motivation. This seems to be the case most of the time. There are fundamental principles that can be applied to nutrition education. Educators must create awareness (get the attention of the intended audience), provide motivation to accept or seek the message, and give the message.

The message must be given at the proper emotional level, through appropriate media, and should be designed to bring about behavioral or value change. The use of a positive or negative approach (benefit vs harm) will depend upon the audience and the message content. Initial information must be minimal and capable of being related to a continuum of information later. Reinforcement of the message may be desirable.

The concept of the basic food groups around which so much of nutrition education revolves has not been kept up-to-date. It was valid for the staple foods *circa* 1955 and for food items that can be easily classified into one group or another. The food groups are not really appropriate for many of the new foods.

The eager learner can be taught how to figuratively "dissect" recipes or mixed dishes and apportion the ingredients to the appropriate food group. Perhaps more attention should be given to this aspect of education in the case of the concept of food grouping. But, until the consumer has been motivated to become concerned about food combinations and has grasped the concept that foods are the preferred sources of nutrients, no amount of teaching aids will be of value.

The limitations of the label as a medium of consumer education must be understood. Too often it is suggested that the consumer ought to be able to find what he needs and what he wants to know from the label. The educator sees the label as a means of providing useful and meaningful information. Perhaps so, but the label does not and probably cannot provide complete information. For example, labels on products enriched according to standards of identity need not include the list of ingredients. Seldom does a label inform the consumer how to combine the food item with others to make a nourishing meal. An industry panel of the White House Conference on Food, Nutrition, and Health (WHC) stated the requisites for informative labeling: "There are three reasons for uniform labeling of all processed foods:

"1. To identify the product and tell the consumer how to use it effectively.

"2. To inform the consumer of ingredients or properties which may be of significance in terms of particular nutritional or health needs.

"3. To permit the food processor to

advertise his product to the public in a competitive market." (WHC Panel III-4—portion of Recommendation 2, page 143).

Clearly, consumer education programs will be needed to assist in the proper utilization of nutrition labeling. It is difficult to conceive how a nutrition education program could be devised with its sole basis being nutrient information on a food label. Such information could assist the consumer who wants to be assured that her meals are nourishing.

One possibility based upon the concept of four food groups (classes) would be to publish average values of nutrient composition for classes of foods. The consumer then could compare the label of a member of that class to determine if it generally fits into that class or if some other food should serve as a substitute. This relates a given food to a well-tested plan for the combination of foods into a day's nourishing intake.

Food advertising and nutrition education too often are at odds. Similarities in mission exist—but the function of each differs and should be understood. Since each is an integral part of consumer education, it is only reasonable to expect that education and labeling should be complementary. An exploration of basic differences might help our understanding. Food advertising nearly always extolls the virtues of a given food. That food is seldom placed into the context of a whole meal. The consumer is seldom informed how to use the food in combination with others to make a nourishing meal. Advertising of fun foods or snacks is too often counter productive to the efforts of the nutrition educator.

Advertising is motivation-oriented while education is not oriented to specific decision making, but rather emphasizes free choice. Advertising is finite while education is open ended. If the two are ever to work together, there must be some common understanding; the responsibility of the educator is to "transmit" a "body of knowledge." The responsibility of the advertiser would be to motivate consumers to act responsibly within that body of knowledge. This, for the most part, was the case during the consumer education phase of the enrichment movement in the 1940s. But those days are gone forever.

Certain practices and trends in food product design, education, promotion, and/or advertising mitigate against programs designed to encourage the use of enriched or fortified foods. The most obvious of these is that fortified foods not covered by a concept of standards reduce the effectiveness of the key words—fortified and enriched. Even the word "restored" now has little significance.

Ideally, the use of these words should be reserved for the identification of foods of special significance. Otherwise, the consumer has no way of knowing if fortification has significantly improved the nutritional quality of the food item. As experience is gained with nutritional labeling, there exists the possibility that processors will add nutrients to foods (when not in conflict with a Standard of Identity) in order to permit their listing on the label. Such practices could further confuse the consumer who is trying to comprehend the significance of the key words—enriched and fortified.

In the absence of "before and after" studies of the benefit of nutrient enhancement of foods, the educator does not have a strong basis for recommendations and teaching. Information is not available with which to judge the effectiveness of fortification programs, which is necessary for continued educational support. The original concepts of enrichment and fortification were developed a generation ago when examples of classical nutrient deficiencies were relatively common. Such is no longer the case. Nutrient enhancement of foods is now a supportive measure rather than one of correction of dietary inadequacies. Thus, arguments in support of enrichment and fortification must stem from a different base.

Just as the basis for nutrient enhancement changed, so has the basis for Standards of Identity of foods. This is significant because for some foods the nutrient enhancement is written into the standard (eg, enriched bread) or prohibited when provision was not made for same. The older standards were recipe-type descriptions of foods to assure reasonable similarity of identically labeled products. Current thinking is that standards of identity would better be written around the valuable ingredient(s) giving more latitude to product design. The FDA recently encouraged processors to list the ingredients in standardized foods as a means to better inform the consumer. Improved labeling would also assist the consumer to a better understanding of the terms *enriched, fortified,* and *restored.*

When the consumer is urged to consume a variety of foods (selecting from classes of foods), he should expect that the foods will be nourishing, that is, that they will provide a proportionate quantity of the nutrients. If this is an acceptable axiom, then a food item that does not provide its apportioned share of nutrients should be altered or the general recommendations should be clarified. Evaluation of the usefulness of this axiom requires the examination of current concepts of nutrition education and/or review of food composition.

The practices and trends just reviewed lead to the question, "Is it reasonable to have on the market both enriched and nonenriched, otherwise identical, products? That is, when a food item is selected for nutritional improvement, should improvement be optional?" While it is recognized that great effort would be required to change standards of identity promulgated by the various states, such effort would be worthwhile. Clearly, the obligation of the educator is to recommend use of the improved food when there is compelling reason to fortify.

CAN CONSUMER EDUCATION SUBSTITUTE FOR FOOD FORTIFICATION

Programs of food fortification probably cannot compensate for really poor food habits. Clearly, to benefit, people should consume the fortified products or else enjoy a varied diet that will provide the needed nutrients naturally. For fortification to work, there must be some consistency in food habits (requiring extensive education). There will always be some who eat as they please and cannot be reached by conventional means of consumer education. Education instead of fortification may have limited usefulness for the majority of people. This can be illustrated by an apt example:

The consumption of dairy products (excluding butter) in 1958 was 672 pounds per person. By 1970, consumption had declined 102.4 pounds per person per year; the projection to 1980 predicts a further reduction of another 122 pounds to 447 pounds per person per year. This predicts a 33 percent decline in the consumption of dairy products in 20 years.

Could education be expected to reverse this trend? It is unlikely, since dairy products are heavily promoted in consumer education and advertising efforts. But what about the resulting decline in consumption of calcium, riboflavin, and other nutrients provided by milk? The slack will have to be taken up by fortification and/or the development of appropriate substitute sources.

It is doubtful that education could reverse the trend toward convenience—or substitute—foods of lower nutrient density. When this occurs at the expense of traditional or staple foods, appropriate fortification is necessary. Education alone could not be expected to obviate the need for fortification under these circumstances.

When nutrient balance has gone completely out of kilter, it is too late for consumer education to correct the situation. While the food supply is being righted or food habits are being restructured, fortification of appropriate foods is almost mandatory. If anything, the educational thrust must be redirected in order to impress and inform the industry and policy makers. It seems reasonable to conclude that consumer education cannot be an appropriate substitute for food fortification.

CONSUMER EDUCATION AS AN ADJUNCT TO FORTIFICATION

Sanctioned programs of food fortification should include multi-faceted educational efforts. Consumers should be encouraged and taught how to use the fortified product. There should be few exceptions to the fortification policy. When a given good is selected for a nutrient enhancement by fortification, every effort should be made to have universal regulations covering that food. If education is to be a useful adjunct to fortification, fortification must follow logical and recognizable guidelines.

Consistency of terms is of utmost importance. The term *enriched* is generally reserved for use with cereal products (flour, bread, rolls, macaroni, corn meal, and rice). The term means fortified with more than one nutrient in conformity with a government standard of identity. *Fortified* means the addition of one or more nutrients that were not present or were found in lesser amounts before processing. At the present time, few if any standards regulate fortified foods.

When the producer is permitted to make claims for the fortified food items, ie, call attention to the added nutrients, he will help carry out the educational program. There is special provision for this under the new food labeling regulations.

A national center for coordination of nutrition education would do much to upgrade the quality and extent of nutrition education. The Inter-Agency Committee on Nutrition Education could be effective if the member agencies demonstrated sufficient interest. Until that happens, we must look elsewhere for stimulation.

When I look back over the more than 30 years since pure vitamins and minerals and eventually amino acids first became commercially available at prices that made it feasible to enrich and fortify foods, I began to realize that many problems have been encountered. Many of these problems have been solved but several, especially those involving the use of iron and iron compounds, are still awaiting the right answers.

Improvements have been made in the enrichment formula for cereal products. Some years ago, we changed from thiamine hydrochloride to thiamine mononitrate, thus improving the stability of this important ingredient.

In the United States, there have been problems with iron compounds. I have become keenly aware of the need for right answers in the developing countries, especially in India where the US Agency for International Development has collaborated with the Indian Food Ministry in a project for fortification of solar salt with calcium, iron, phosphate, and other nutrients. We thought we might be able to incorporate lysine in salt. We spent some time on these studies. We found we could, to be sure, add lysine to salt; but after a few weeks or a few months, the salt developed an objectionable odor and somewhat to our surprise it became interesting to insects and actually became infested.

The development work on salt is practically at a standstill. We had started out thinking that we could use insoluble iron salts which would not discolor the sodium chloride. But discredit was thrown on these insoluble iron salts by experiments by Indian nutritionists with a very small number of human subjects, which seemed to show that the salts were almost completely unavailable.

There are some criteria for fortification which might be worthy of mention in view of our years of experience in the United States with fortification of various foods, especially the cereal foods. The stability of the vitamins and minerals to be incorporated in the food under consideration should be carefully studied under the anticipated storage conditions for the length of time the food is to be stored.

The effect of fortification ingredients on the organoleptic properties of the food must be examined both initially and during storage. Here, again, this especially refers to

chapter 15
manufacturing problems in fortifying foods

by
Clinton L. Brooke

certain of the iron compounds, particularly ferrous sulfate. Thiamine because of its odor has to be watched carefully and the level must not be too high. The same is true of niacin.

The effects of baking at the prescribed temperature and time on the color, flavor, and vitamin levels of baked products must be determined by inspection and by assay. Appropriate corrections must be made for processing losses. The problem of maintaining thiamine levels in alkaline cakes, such as devil's food cake, is a troublesome one and it has yet to be solved. It may be solved through encapsulation of the enrichment ingedients.

In the manufacturing of breakfast cereals, consideration must be given to the stability of the added nutrients. Those that are not heat resistant are often applied to the finished cereal product immediately

141

after toasting. Heat-stable nutrients may be incorporated as far back as the cooking of the grits or whatever the raw material may be. Then we have the continual problems of iron compounds in foods, and the question of assimilability of the iron compounds.

Ferrous sulfate is probably the most acceptable nutritionally. Recent work by the FDA gave hopeful indications that increased levels of the less soluble iron salts could provide levels of hemoglobin regeneration equivalent to those with lower levels of ferrous sulfate. This, incidentally, has been presented to Indian nutritionists but they are not willing to accept it as yet. They are still asking for further studies on human subjects. In flour and cake mixes, ferrous sulfate may cause discoloration as well as changes in flavor.

Rice enrichment problems are not entirely solved. Powdered vitamin mixtures are now being added to much of the packaged rice produced in this country, and premixes consisting of rice kernels covered with vitamins and iron and protected with a rinse-resistant coating have been used in the United States, especially in Puerto Rico. The optional formula for rice fortification in the United States does not include riboflavin because riboflavin imparts a yellow color to cooked rice.

The development of stable dry forms of vitamin A has greatly broadened the scope for vitamin A fortification. I had my first contact with this dry form of vitamin A in Jordan in 1966. After examining the possibility of adding vitamin A to olive oil, and discarding it because olive oil is not well enough distributed throughout the Jordanian population, we decided to add vitamin A to flour. This was the first time dry vitamin A had been added to flour in a flour mill. We used a mixture of the usual enrichment ingredients plus vitamin A in beadlet form.

These were rather coarse beadlets, about 60 to 80 mesh, and they worked out very well, but when we took a recommendation for vitamin A fortification of flour back to the United States, the flour mills said they could not use vitamin A beadlets of this particular size without changing the sieves at the tail of the mill. Industry rose to the emergency and induced a leading manufacturer of vitamin A to produce a suitable form which could be added to flour and would not be sifted out at the tail of the mill.

This type of vitamin A, namely, stabilized vitamin A palmitate, has made it possible to fortify tea. We're not yet sure exactly at what point it will be added but we are sure, from studies of the technology of vitamin A fortification of tea now (1973) in progress in the United States and also in Bombay, in the laboratories of this same vitamin manufacturer, that this can be done. The infusion of tea fortified with vitamin A can be boiled for as long as one hour with no loss of potency. It will still be necessary to conduct clinical studies to see if there's any loss of vitamin A potency in the cooking of the tea.

Proposed changes in the enrichment levels for flour, bread, corn meals, rice, etc., will require changes in enrichment mixtures for the miller and in enrichment tablets for the baker. Manufacturers of enrichment products have given assurance that they can meet the changes in formula.

One of the overwhelming and inescapable conclusions from all the research data we have seen is that *in the eyes of the consumers* there is no significant problem about nutrition. While there are some obvious exceptions for certain segments of consumers and for all consumers under specific circumstances, most people are satisfied that most of the time they are doing a good job of satisfying the nutritional needs of their families. They are completely aware that good nutrition is important and that it would be bad if they failed to meet the nutritional needs of their families.

In survey after survey, in response to wide ranges of research approaches to the issue, 80 to 90 percent of the respondents report that their families obtain nutritionally adequate meals, that their families are healthy, and that their eating habits (however distressing meal skipping and snacking may be) provide adequate nutrition. The only recurring problem with respect to the broad area of food and health, as consumers define problems, is in the area of overweight and dieting. About 50 percent or more express concern about overweight and *claim* some form of overt activity to deal with that problem.

There appear to be two principal bases for the consumers' confidence in the nutritional adequacy of their present diets and eating habits. The first of these is a very common-sense functional test. If their families, particularly the children, are well, active, and look good, with good color and bright eyes, the homemaker has confidence that she has done a good job of feeding the family what is needed. The second basis for confidence is the homemakers' reliance on a very loose, generalized idea of "balancing."

Most homemakers consistently believe that if they serve their families enough food and if, over a reasonable period of time, such as a week, they manage to introduce into the diet an adequate variety of meat, potatoes, vegetables, fruits, and milk, they will have met the need for nutritional balance. In the homemakers' eyes, the principal place and time for achieving adequate "balance" are found in the treatment of the evening meal. This is generally the most carefully planned meal of the day.

"Balancing" is done on the basis of "folklore" about nutrition. The benefits that could be obtained by the proper

chapter 16
problems of researching and marketing fortified foods and their implications for consumption trends

by
**Howard E. Bauman
PhD.
and Dudley Ruch**

balance of nutrients by meal and by day are not very well understood. In a number of studies, consumers were asked to describe nutritionally balanced meals or menus. Only about 50 percent came reasonably close. Thirty to 40 percent left out one or more of the basic foods that should be in a menu. In these studies, the question was not on the *amount* that should be consumed, only the *identification of what.*

In measuring nutritional knowledge, one sample was asked to name the nutrients contained in 14 foods and another sample was asked to tell us the foods that would be important sources for 7 common nutrients, expressed in layman's terms. For the 14 foods, a maximum knowledge score of 31 was possible; the average score was 11. For the nutrients, the average number of foods named was 13 out of more than 100 common food sources of these nutrients. The functional knowledge about what foods contain which nutrients and about what benefits are derived *either* from the nutrients or from specified foods is *spotty*. The things that are best known are:

1. Calcium comes from milk and builds bones and teeth. When asked the principal benefits of milk, however, consumers ranked vitamins first. There is a low level of knowledge of milk as a source of protein or of benefits associated with milk protein.

2. Meats and poultry provide protein which builds muscles and provides strength.

3. Orange juice provides vitamins which prevent colds.

4. Sugar provides quick energy—few know it is a carbohydrate; it's fattening and bad for the teeth.

5. Iron builds blood; it comes from liver—other sources are not known.

6. Potatoes and baked goods have the same image; nutritional benefit is primarily to be filling—little knowledge of true nutritional benefits and high concern with being fattening.

7. Fats in the form of most dairy products, except milk, have no clear nutritional benefit, are bad for you because they are fattening and contribute to heart disease.

8. Breakfast cereals provide vitamins which make them healthy.

9. Vitamins are good for you in a vague way—only specific benefits are orange juice and cold prevention, and carrots and vision.

There is a serious gap between attitudes and perceptions about nutrition and those about weight control. All the common sense about balancing goes out the window when the subject of weight control comes in. There is very little awareness or concern about the nutritional impact of various weight reduction or control techniques—breakfast or lunch skipping, total omission of certain foods from the diet, or fad diets. A factor which should greatly concern us is that reliance is placed on the image of some kinds of foods being fattening, and therefore having no important nutritional value—*potatoes; breads and pastries; sugar; dairy products such as cream, cheese, butter and ice cream; and snacks.*

Because of the "avoidance" character of weight control, calories are understood only in the context of contributing to overweight. There is no understanding of calories as contributors to energy or of the effect on nutritional balance of calories from different food sources. *All* calories from any source are bad. There are *some* exceptions to this general conclusion about nutrition and weight control which have to do with:

1. Mothers of teenage daughters where dieting practices go counter to what a mother believes a "child" should be eating. *But the same mothers will often follow the very dieting practices they object to for their daughters.*

2. Middle-aged and older, upper-class wives who are seriously concerned about their husband's diets with respect to overweight and potential heart disease. These women have a two-way conflict—about what he eats at noon, or when he travels, and about his resistance to her efforts to move away from heavy meat and potato meals at dinner.

Given a general level of satisfaction with the nutritional adequacy of her family's diet and a superficial, but not totally inaccurate level of nutritional knowledge, the homemaker's primary concern with food is in pleasing her family. The principal motivating forces that have to do with the selection of foods to be admitted to the family diet are first, *sensory gratification;* second, *accommodating the likes and dislikes of individual members of the family* (in most homes, the husband's desires about food place outer boundaries on what will be served); and third, accommodating to the increasingly difficult problem of meal scheduling.

Every piece of research that we have done or have seen done by someone else emphasizes the paramount importance of family acceptance of the food as tasting good and as being the kind of food they want to eat. A sizable minority of women struggle valiantly to get their husbands to go beyond "plain foods"—meat, potatoes, and vegetables—to wider ranges of vegetables, salads, and fruits; and they struggle with their children to get them to eat vegetables. *However, in survey after survey, women admit that* when faced with the choice between serving less appetizing nutritional foods and foods that the family likes, even though they are not as nutritional, they usually opt in favor of the appetizing but less nutritional foods.

The problem of individual food likes and dislikes is recognized by women as being one of the biggest single barriers to their doing a better nutritional job (in the context of their belief that they're doing a good job now). The principal issues are:

1. With children over inadequate consumption of milk and vegetables.
2. *Children's propensity to overconsume snacks and soft drinks which can spoil the children's appetite for the dinner meal which is intended to relieve the housewife's obligation toward nutritional balance.*
3. *The limitations of the husband's traditionally narrow range of food interest.*

The latter is a continuing source of frustration to a number of women along two dimensions: first, and most important, her desire to show her skill as a cook in trying new things; and second, the fact that the narrowness of her husband's range reduces her ability to get him to eat various kinds of vegetables and salads and fruits that she feels would contribute to more favorable nutritional balance.

The accelerating disintegration of the old, traditional pattern of the family being together for three well-defined meal occasions a day has a direct effect on homemakers' attitudes and concerns about nutrition. There is also a rather significant gap between consumers' concerns and beliefs about how well they do their nutritional job and their behavior.

There is a significant amount of meal skipping—particularly breakfast and lunch. *Thirty to 50 percent of families have one or more members who fairly regularly skip breakfast. Three-fourths of the families do not eat breakfast as a family unit* and in about one-third of the families, the mother has no part of serving or preparing breakfast for any member of the family.

As a result of breakfast eating/ skipping patterns, *no less than 50 percent of school age children have either no breakfast or a nutritionally inadequate breakfast.*

In our analysis of the nutritional adequacy of breakfast based on the *MRCA menu census, we found that about 85 percent of all breakfasts served provide less than 25 percent of the total daily*

calories. More than 50 percent of the breakfasts furnished less than 25 percent of the protein RDA. While we don't have good detailed studies on lunch eating behavior, some of the studies we have would suggest *as high as 25 or 30 percent of the women who are at home alone skip lunch.* Two major themes emerge as important contributors to nutritionally inadequate eating habits:

1. Dieting and weight control result in skipping or skimping breakfast and/or lunch.
2. The breakdown of the traditional three meals a day pattern of eating behavior.

Of extreme importance concerning the breakdown of the traditional three-meal, together-eating pattern is that *there is no convenient way to connect the nutritional knowledge a woman has to the newly emerging eating patterns and foods. Most nutritional education, in the schools and in the mass media, is based on three traditional meals a day, and the basic seven. The housewife is at a total loss about meeting nutritional needs in the reality of today's individualized eating patterns with foods that fit these patterns.*

Of even more concern is that none of our studies shows any *direct consumer concern* with how changing eating patterns adversely affect nutrition. *The indirect concerns are reflected in the negative connotations of snack foods and soft drinks rather than a positive awareness or concern for missing nutrients.*

While there are exceptions that show consumption trends and nutritional attitudes going in the same direction, *consumer behavior clearly supports the idea that the hedonistic and social values of food are far more relevant to consumption decisions than nutritional issues.* In our Nutritional Base Line Study, we attempted to segment the population on the basis of nutritional attitudes, behavior, response to nutritional product concepts, and demographic characteristics. From this analysis, we could distinguish some groups of consumers with respect to these variables and make approximate estimates of their size.

The following is a necessarily over-simplified combination of the 11 nutritional attitude segments found in the Nutritional Base Line Study. The borderlines between the groups are grey. The descriptions of the groups are based on average tendencies for the purpose of describing the dynamics of how demographics, nutritional attitudes, and nutritional behavior interact.

Group I represents *10 to 15 percent of housewives* and includes many of the families with the highest income and education in the sample. They're above average in nutritional knowledge. They are below average in response to nutritional product concepts as measured by nine point mean ratings and buying intention scales. These women have the fewest problems in meeting their families' nutritional needs and are most able to cope with the kinds of nutritional problems that concern women in other groups—and are thus least interested in special products to meet nutritional needs.

They have largely solved the problem of weight control and calorie counting and are below average in their concern for dieting. They have a high concern for nutritional balance and might be expected to respond to informative communications on labels and in advertising about nutritional balance and intelligent weight control aspects of foods.

Group II consists of about *30 percent of the consumers.* Traditional values about homemaking put them in the upper half of the population with respect to nutritional attitudes and behavior. Most are above average in socioeconomic characteristics—with the remainder consisting of older homemakers with very traditional views about meal scheduling and proper eating.

Nutritional knowledge is average. However, their attitudes (traditional) lead to average or above average nutritional adequacy of meals. These are the women to whom nutritional and weight control information and the editorial material of the type found in magazines—such as *Better Homes and Gardens* and *Good Housekeeping*—would have the greatest appeal.

Group III contains *25 to 30 percent of the housewives,* and they are fussy eaters whose nutritional attitudes and behavior are overshadowed by their own likes and dislikes for specific foods. They admit to a tendency to ignore their own diets and their families are heavy consumers of food. They tend to be below average in their concern for nutrition and less con-

cerned about the problems of pleasing their family with nutritional foods.

Group IV represents 25 to 30 percent of the housewives and is the bottom group in nutritional knowledge, attitudes, and behavior. This group is well below average in income and education. Meal scheduling which is related to a below average feeling of family cohesiveness is a real problem as reported by these women. The younger women are described as "detached" with little involvement in cooking while the older women are apt to be working wives, which contributes to the meal scheduling problems.

In summary, what this analysis market segmentation shows is that the people who have the most knowledge and interest in nutrition are doing a good job in providing nutritionally adequate meals for their families and have the lowest level of interest in nutritional products. At the other extreme, the groups that are most in need of improved nutrition have the least knowledge and concern about the subject. These people need solutions to problems they already recognize—*meal scheduling, fussy eaters,* nutritional snack foods that aren't fattening, etc. For these people, nutrition claims are like vitamins—a reinforcement in the sense that you can't have too much of it, but that is not a primary reason to buy.

A major problem is communication, especially as it may relate to the snack world. The data show that the promise of sensory reward is the crucial benefit for any food product. Only after that promise is adequately communicated in advertising or on labels will nutritional claims be considered by most consumers. If the "good tasting" communication is inadequate, nutritional claims will appeal only to very small group actively concerned with nutrition.

In trying to trace meal replacement products that have been placed on the market in the last few years, we can look at instant breakfast. We find that instant breakfast is not replacing any type of traditional meal; it appeals primarily to women and teenagers who are breakfast skippers anyway. Generally, it substitutes for coffee and toast.

As far as snack foods go, the largest area are cookies and crackers. Sales of these items are about $1.4 billion, representing about 50 percent of the total snack food market as it's measured today. This area probably represents the biggest change in fortification, since most of the crackers and cookies on the market now are either fortified or in the process of being fortified; so, we can say that at least 50 percent of the snack foods have had a considerable nutritional change.

The next largest—about 20 percent of the market—is potato chips—none of these as yet are fortified or restored. Corn chips and pretzels each have about 4 percent of the market and the likelihood that these will be enriched is good. Nut-meats are about 15 percent of the market and do provide a fair amount of nutrients. "Other snacks"—the extruded, expanded, fat-fried types—have about 4 percent of the market and, since most of these are cereal-based, will most likely be enriched, but are not as yet. Meat snacks are about $60 million in sales—mainly jerky and sausage types; toast'n pastries are about $60 million and are enriched; and frozen, pre-packaged pizza has a market of about $145 million. In the past 10 years, candy consumption has gone down in spite of an increase in population—down about four percent from what it was five years ago. But soft drink consumption has increased. In 1940, the consumption was 100 8-oz bottles per capita. In 1950, it was 158 per capita. In 1960, it was 192 bottles per capita; and in 1968, it went to 332 bottles per capita—getting close to one bottle per day for every man, woman, and child in the United States.

The total market for snack foods is about $3.5 billion and the indications are that all of the markets with the exception of candy are growing from 4 to 10 percent a year. We'll see a lot more new types of snack foods coming in because the whole trend of eating is zooming toward the 5, 6, 7 or 8 smaller meals a day, with one large meal—which is usually the evening meal—jammed in. Since snacks do constitute a significant number of calories in the diet, it seems the next step should be to determine what they generally replace and these factors should be taken into account when guidelines are prepared.

At the 1971 symposium on "Vitamins and Minerals in Processed Foods,"[1] discussion was centered on how knowledge gained from insuring nutritional adequacy of foods for "special dietary uses" can be applied to improving the nutritional value of processed foods in the ordinary diet. The following nutritional concepts regarding the addition of nutrients to foods for the general public and problems of the federal regulations in effect at the time were emphasized at the symposium:

Fortification of processed foods that replace individual conventional foods or mixtures of conventional foods in the diet was strongly advocated but only with nutritionally meaningful combinations and amounts of nutrients—widespread and haphazard fortification of conventional foods was not endorsed.

Products on the market at that time varied considerably in their content of essential nutrients and it was almost impossible for consumers to know whether these products contained proper combinations and amounts of essential nutrients for their roles in the diet—or to tell whether advertising statements for these were nutritionally sound.

Regulations which defined foods for "special dietary uses"—those products used in the dietary management of persons during disease and convalescence, and products used as the sole item of diet—and dealt with labeling of these foods needed to be redefined so that they applied only to those foods which really had "special dietary uses." Under the regulations existing at that time, addition of a nutrient to any conventional food automatically made it a food for "special dietary use" even when the food was intended for consumption by the general public.

The only nutrient labeling in use was based on somewhat obsolescent Minimum Daily Requirements (MDRs), which had been set for only a few vitamins and minerals essential to man; appropriate Recommended Dietary Allowances (RDAs) were needed in the regulations for all nutrients essential to man so that they could be properly applied to fortification and labeling of both special and regular foods.

Since the 1971 symposium, the US Food and Drug Administration has effected changes in the old nutritional regulations along some of the lines suggested above

chapter 17
fortification of foods for general use vs those for "special dietary uses"

by
Herbert P. Sarett, PhD

and also has adopted some new regulations and guidelines. On Jan. 19, 1973, proposed regulations were issued in the *Federal Register*[2] for: §80.1—which governs vitamin and mineral supplements; §125—which concerns foods for "special dietary uses"; and §1.17—a new section to regulate nutritional labeling of regular foods. Most of these sections have now been finalized—with some further modifications, as published in the *Federal Register* of Mar. 14,[3] and Aug. 2, 1973.[4]

Discussions of nutritional regulatory problems at the New Orleans symposium, as well as at others,[1,5,6,7] have had a beneficial effect on some parts of the final regulations. It is pertinent at this time to review the manner in which foods were fortified with nutrients on the basis of the old regulations and to determine to what extent the new regulations more effectively allow or control addition of nutrients to special and regular foods.

For many years after the 1941 regulations[8] were introduced, foods for "special dietary uses" were mainly those used as infant formulas and those designed to meet particular dietary needs during conditions of disease and convalescence. Since the 1941 regulations applied to all foods to which nutrients were added, an increasing number of foods that were designed and formulated in recent years for the general public—ie, breakfast cereals, instant breakfasts, low-calorie foods, meal replacements, etc.—had to be labeled as foods for "special dietary uses."

Even nutritional labeling proposals by the FDA in 1971 and 1972 for other classes of conventional foods continued this practice.[9,10,11] As a result of many comments on this problem, the FDA has issued a new section, §1.17—which provides for nutrient labeling of regular foods—and has redefined foods for "special dietary uses" under §125 as being those foods which are *used* to supply special dietary needs.

MDRs AND RDAs

The 1941 regulations included Minimum Daily Requirements (MDRs) for only a few of the nutrients now recognized as essential to man. A manufacturer who added or labeled a food product with respect to only seven nutrients could legally state that the food "supplied all of the officially established Minimum Daily Requirements for vitamins and iron."

This type of promotional statement often grossly misled American consumers and even some members of the medical profession into thinking that various fortified foods and vitamin and iron supplements provided *all* of the nutrients needed in the diet. The nutrients for which there are established RDAs and others considered essential for man are listed in Table 41.

For labeling purposes, MDRs had been set for only vitamin A, vitamin D, ascorbic acid, thiamine, riboflavin, niacin, calcium, phosphorus, iron, and iodine. Attention was paid mainly to MDRs in foods for general use in the diet, as well as in many vitamin and mineral supplements.

Even foods professing to be fairly complete included only those nutrients with established RDAs and not necessarily all nutrients recognized as essential. It was only in foods that really had "special dietary uses" that most of the nutrients listed in Table 41 were included—depending on the type of special use and the judgment of the manufacturer.

FOODS FOR "SPECIAL DIETARY USES"

Really "special" foods such as infant formulas, tube feedings, and sole or major items of diet had to be nutritionally complete if they were to support good growth of infants or maintain adult subjects in good nutritional status over extended periods of time. (Nutrient content of these products will not be significantly changed by the new regulations.) Table 42 shows the nutrient content of two infant formulas and two other special dietary foods that included nutritionally significant amounts of practically all nutrients discussed and listed in Table 41.

The infant formulas shown are milk-substitute formulas made from isolated and/or purified ingredients for infants who cannot tolerate milk. The protein in the first formula is provided by isolated soy protein supplemented with methionine and in the second by a protein hydrolysate. All essential nutrients are added to these formulas because:

1. The purified protein, fat, and carbohydrates do not concomitantly provide other nutrients.
2. The formula may be the sole item of diet for an extended period of time.
3. Infants consuming these formulas may receive only a limited selection of supplementary foods due to allergy or illness.

The MCT product for adults and children in Table 42 contains medium chain triglycerides as the main source of fat, is free of lactose, and is intended for persons with malabsorption. It is formulated to supply reasonably adequate levels of all essential vitamins and minerals for man in 960 Kcal so that the remainder of the diet can be chosen on the basis of patient acceptability, rather than with marked emphasis on nutritional value. At higher caloric intakes, the product can also serve as the sole item of diet.

The tube feeding product in Table 42 also supplies all essential nutrients. A day's requirements are included in 1,080 Kcal of this product, since many bedridden subjects are kept on low caloric intakes; persons who consume 1,500 to 2,000 Kcal of the product per day receive 1-1½ to 2 times the RDAs. In designing a special dietary food such as this, it is more important to assure adequate levels of nutrients for persons receiving smaller amounts of the product than to be overly concerned that individuals requiring higher caloric intakes receive a little more than their RDAs.

MEAL REPLACEMENTS

In recent years, new meal replacement products have been developed and labeled as "special dietary foods" for use by the general public. The first such products were intended for use in weight control regimes and were nutritionally complete (except for calories), well-balanced, and supplied about one-third of the day's requirements of all nutrients per serving.

However, some of the more recent products no longer provide a nutritionally complete or balanced small "meal"—Table 43, particularly when a powder is added to a glass of milk to provide an "instant breakfast." Product *A* (with milk) provides only trace amounts of vitamin E, pantothenic acid, folic acid, iodine, manganese, zinc, and copper. Product *B* is more complete when consumed with milk and most of the nutrients are present at about 25 to 30 percent of adult RDAs; but levels of folic acid, magnesium, and zinc are still quite low for a meal.

If the rest of the day's diet provides a good mixture of conventional foods, these low levels of nutrients in one meal are not important, but the lack of completeness of these products is of concern when similar products or other partially fortified foods provide most of the diet. (Nutritional adequacy of these products will be specified by nutritional guidelines to be proposed by the FDA in the near future.)

BREAKFAST CEREALS

Breakfast cereals also have varied considerably in their nutrient content—Table 44. Some gave the impression to the public that they supplied all the micronutrients that are needed in the daily diet by advertising that they contained the

MDRs of all vitamins (and iron) established by the US government. Products such as *C* and *D* in Table 44 contain the MDRs of thiamine, riboflavin, and niacin—as well as of vitamins A, D, and C which aren't shown in the table—but are essentially devoid of other essential vitamins and minerals needed in the diet; cereals such as *A* and *B* aren't heavily fortified but provide a more balanced array of vitamins and minerals.

It's nutritionally more sound to have moderate levels of all nutrients in 100 Kcal of cereal than to provide the entire day's requirements of a few nutrients and only traces of the others. Since fortified breakfast cereals may be considered a type of meal replacement, it would be best to include all essential nutrients at about 1/4 to 1/3 of the day's needs. It is hoped that FDA's forthcoming nutritional guidelines on breakfast cereals will call for this type of balanced nutrient addition.

IMITATION FOODS

In the past, there have been no legal criteria for including nutrients in imitation foods. FDA proposals, such as those for textured vegetable proteins[12] and imitation milk,[10] have ignored many of the essential nutrients present in the foods that were being imitated. The Joint Statement of the American Medical Association and the Food and Nutrition Board[13] recommends that:

"The imitation or fabricated food should contain on a caloric basis at least the variety and the amounts of the important nutrients contained in the food which it replaces."

On this basis, one should consider the nutrients listed in Table 45 in formulating imitation foods, nutritionally fortified snacks, meal replacements, or any products that play important roles in supplementing or providing part of the diet. Levels of the nutrients can be included on the basis of daily intakes, amounts per meal, or where appropriate in amounts/100 Kcal, as shown in Table 45.

In the new regulations on Imitation Foods—§1.8—proposed on Jan. 19, 1973[14] and finalized on Aug. 2, 1973,[15] the FDA considers most of these nutrients as essential for purposes of imitation foods, but has set unrealistic standards for the level of a nutrient present in a conventional food that must be copied in the imitation food.

153

The January 19 proposal would have allowed imitation foods to be considered nutritionally equivalent if they contained 10 percent *less* of the day's RDAs of nutrients in a serving than the food being copied.

Since single servings of many good foods may contain only 5 to 10 percent of US RDAs, these imitations could have been completely devoid of all essential nutrients. This was pointed out in our letter of Mar. 15, 1973, to the FDA, in which we also proposed that "an essential nutrient should be included in a substitute food if the original food provides at least 5 percent of the US RDA of the nutrient in a serving or in the usual day's intake, and at least 1.5 percent of the US RDA per 100 Kcal."

In the August 2 publication,[15] the FDA recognized the inadequacy of the January 19 proposal, but went on to the other extreme in assuring nutritional equivalence in the final regulations by requiring the presence in the imitation food of protein and of all essential vitamins and minerals present at *2 percent* or more of the US RDA per serving of the food. This regulation is completely impractical and unrealistic since it will require the costly inclusion of nutritionally insignificant trace amounts of many nutrients in imitation foods.

For example, even Mellorine and Parevine, which imitate a 174 Kcal serving of ice cream, would be required to contain thiamine (2.7 percent of US RDA), pantothenic acid (3.9 percent of US RDA), vitamin B_6 (2.5 percent of US RDA), copper (2.0 percent of US RDA), and zinc (3.4 percent of US RDA), in addition to protein, vitamin A, riboflavin, vitamin B_{12}, calcium, and phosphorus, which are present at from 7.7 to 13.1 percent of the US RDAs.

Foods with higher caloric levels per serving would be even more seriously affected by this regulation. If standards for textured vegetable protein, imitation milk, Mellorine, or Parevine were set on this basis, they would require inclusion of a far greater number of nutrients than the FDA appears to realize.

ENRICHED FLOUR

The FDA has done nothing to improve or update the enrichment of flour since 1942 beyond recently proposing an increase in levels of iron, as well as thiamine, riboflavin, and niacin.[16] Many nutritionists in the US have long urged that other vitamins and minerals lost in milling of flour be restored or partially restored to this food that supplies about 500 Kcal/day in the average diet.

At present, enriched flour does not provide adequate amounts/100 Kcal of vitamin B_6, folic acid, pantothenic acid, vitamin E, magnesium, copper, zinc, and manganese—although whole wheat flour is a significant source of all of these nutrients. If enriched flour is supposed to be a nutritionally equivalent "imitation" of whole wheat flour, its nutrient content should be fully reconsidered. It's difficult to understand how the FDA set unnecessarily strict new standards for "Imitation Foods" in §1.8 as discussed above, while allowing (and requiring) a basic food such as enriched flour to contain *only* four of the many essential nutrients present at significant levels in whole wheat flour.

NUTRITIONAL GUIDELINES

The first set of nutritional quality guidelines has been proposed and finalized by the FDA for frozen "heat-and-serve" dinners made from conventional foods.[11,17] Nutritional quality guidelines for a defined class of foods are intended primarily to help consumers identify those products that meet appropriate nutritional standards.

An NAS/NRC committee report[18] that was used as the basis of this guideline proposal showed that individual nutrient levels of such frozen "heat-and-serve" dinners made from meat, poultry, fish, or cheese varied considerably depending on:

1. Which of these is used as the major source of protein.
2. The choice of vegetable.
3. The type of potato or cereal-based product used as the third component.

However, only a *single set of guidelines* was drawn up for *all* "dinners," and these called for certain levels of protein and of only five of the many nutrients that are present at significant levels in "dinners." Thus, many nutritionally sound "dinners" that are prepared properly—and contain nutrients in amounts similar to those reported for the same combinations of foods according to Bowes and Church[19] or *Handbook 8*[20]—do not meet the guideline standards, only because of inherent lower levels of iron, vitamin A, etc., in the particular combination of foods in the "dinners."

In order to prevent promiscuous addition of the five nutrients in the guideline to those "dinners" made from conventional foods, and to make sure that the "dinners" supply other essential nutrients that should be in the "dinners" (but for which no values were given), the NAS/NRC Committee recommended that addition of only up to two of the guideline nutrients per "dinner" should be allowed. This was in the FDA's proposal; but, in the final guideline, the FDA allowed the addition of the five nutrients covered by the guideline standards. Thus, *any* "dinner" with enough protein can meet these guidelines by merely adding the five required nutrients, while a number of well-processed, perfectly good products, with all nutrient levels appropriate for the specific types of foods used, cannot meet the guidelines (unless unnecessarily fortified).

In effect, the FDA is not protecting the consumer against nutrient losses due to poor processing, but is asking that a perfectly good fish "dinner" provide more riboflavin than fish ever had, and that a good "dinner" based on cheese provide more iron than cheese can supply—to meet the guidelines. Thus, these first guidelines have unwittingly opened the door to unwarranted and imbalanced fortification of a class of foods that should not need fortification when properly prepared and processed.

Since the stated purpose of the nutritional quality guidelines is to tell the consumer which products contain all the nutrients that this class of product should be expected to contain, we believe that nutrient guidelines have been incorrectly applied to frozen "heat-and-serve" dinners. They should apply primarily to those classes of foods that are formulated and fabricated and need to be fortified to play a specific role in replacing a class of conventional foods in the diet.

The guidelines should consider all the essential nutrients that are present in significant amounts in the class of conventional foods. For frozen "heat-and-serve" dinners, which are combinations of conventional foods, it would have been more reasonable to require nutritional labeling, without fortification, than to establish a single artificial guideline for all such products. This latter only promotes unnecessary fortification and does not guarantee that the food provides all the other essential nutrients which that type of food should contain.

VITAMIN AND MINERAL SUPPLEMENTS

Space does not allow discussion of the new regulations on vitamin and mineral supplements or of foods for "special dietary use." These show some improvement over previous inadequate regulations, but scientific inaccuracies in their derivation and in setting of age groups may essentially prevent both normal persons and patients from getting some of the essential nutrients they need in these products.

NUTRITION LABELING

The impact and possible usefulness of the new nutritional labeling regulations for conventional foods is difficult to judge fully. The required nutrient information on the label for protein, calories, and vitamin and mineral content will be costly to the consumer and it is impossible to know whether the consumer will read, understand, or benefit from the information.

Listing levels of the same few nutrients on all classes of foods does not necessarily provide nutritional understanding. It's unfortunate that the new labeling regulation was not integrated with the traditionally accepted methods of recommending various groups or classes of foods in the diet; it would be easier for the consumer to understand if each of these classes were labeled with respect to those nutrients it was supposed to provide in significant amounts.

In summary, the new regulations which have recently been proposed and, in some cases, finalized by the FDA have made some use of the knowledge gained from fortifying foods for "special dietary uses" during the past 30 years. They have incorporated some of the suggestions made at the New Orleans symposium on "Vitamins and Minerals in Processed Foods" and solved some of the problems which were pointed out at that time:

Foods for "special dietary uses" have been properly defined and separated—for regulatory purposes—from foods

155

intended for consumption by the general public. There are problems foreseen in labeling the latter and these will have to be watched carefully.

US RDAs have been set for most essential nutrients to replace the inadequate and archaic MDRs which had been the basis for all nutritional labeling for three decades; however, the new separation into age groups does not fit physiological needs as well as those used for MDRs.

Nutritional guidelines are being established for some new classes of foods and should help the consumer easily identify the nutritionally sound products in these classes. However, the guidelines for frozen "heat-and-serve" dinners endorse and promote haphazard and un-needed addition of nutrients to these products without a sound nutritional basis.

Regulations have been established to control the amounts and combinations of nutrients in dietary supplements of vitamins and minerals; however, separation of nutrients into mandatory and optional categories does not insure inclusion of all essential nutrients in products in which they may be needed.

These new regulations are vast improvements over those which have been in effect until now. They do take some small but significant steps toward broadening the spectrum of nutrients considered essential in nutrition and in informing the consumer of levels of some nutrients in the foods he purchases.

references

1. Symposium on Vitamins and Minerals in Processed Foods. New Orleans, La., Mar. 22-24, 1971.

2. *Federal Register.* 38, No. 13, 2124-2164, Jan. 19, 1973.

3. *Federal Register.* 38, No. 49, 6950-6975, Mar. 14, 1973.

4. *Federal Register.* 38, No. 148, 20702-20750, Aug. 2, 1973.

5. Sarett, H. P. "Industry's Experience with Special Dietary Foods." *Food Drug Cosmetic Law Journal.* 28: 64, January 1973.

6. Sarett, H. P. "Scientific Issues." *Food Drug Cosmetic Law Journal.* 28: 577, September 1973.

7. Sarett, H. P. "Effectiveness Issues for Vitamins Raises the Question: Are They Foods or Drugs?" *Food Product Development.* 7: 28, June 1973.

8. *Federal Register.* 6, No. 227, 5921-5926, Nov. 22, 1941.

9. *Federal Register.* 37, No. 62, 6197, May 30, 1972.

10. *Federal Register.* 37, No. 176, 18392-18396, Sept. 9, 1972.

11. *Federal Register.* 36, No. 247, 24822-24824, Dec. 23, 1971.

12. *Federal Register.* 35, No. 236, 18530-18531, Dec. 5, 1970.

13. "General Policies in Regard to Improvement of Nutritive Quality of Foods." *JAMA.* 205: 868, 1968.

14. *Federal Register.* 38, No. 13, 2138-2139, Jan. 19, 1973.

15. *Federal Register.* 38, No. 148, 20702-20704, Aug. 2, 1973.

16. *Federal Register.* 36, No. 233, 23074-23076, Dec. 3, 1971.

17. *Federal Register.* 38, No. 49, 6969-6974, Mar. 14, 1973.

18. NAS/NRC Food and Nutrition Board. *Nutritional Guideline Recommendation for Frozen Convenience Dinners.* 1971.

19. Church, C. F. and Church, H. N. *Food Values of Portions Commonly Used: Bowes and Church.* Lippincott, Philadelphia, 1970.

20. Watt, B. K. and Merrill, A. L. *Composition of Foods—Agriculture Handbook 8.* 1963.

Table 41.
Nutrients Essential for Man.

Protein (essential amino acids)	RDA = 65 g
Fat (essential fatty acids)	
Carbohydrate	
Total kilocalories	
Water	

Established RDAs for Man (males, age 22-35, except iron and vitamin D)[*]

Vitamins	RDA	Minerals	RDA
Vitamin A	5,000 IU	Calcium	800 mg
Vitamin D	400 IU (children)	Phosphorus	800 mg
Vitamin E	30 IU	Magnesium	350 mg
Ascorbic acid	60 mg	Iron	18 mg (females)
Thiamine	1.4 mg	Iodine	140 μg
Riboflavin	1.7 mg		
Niacin	18 mg equiv.		
Vitamin B_6	2 mg		
Vitamin B_{12}	5 μg		
Folic acid	0.4 mg		

Other Essential Vitamins & Minerals (estimated RDAs)

Pantothenic acid	(10 mg)	Zinc	(10-15 mg)
Biotin	(150 μg ?)	Copper	(2 mg)
Vitamin K	(?)	Manganese	(4 mg)
Choline	(?)	Sodium	(1,000 mg ?)
Inositol	(?)	Potassium	(2,000 mg ?)
		Chloride	(?)
		Flouride	(1 mg ?)
		Chromium, Cobalt, Molybdenum, Selenium	

[*]NAS Food and Nutrition Board Recommended Dietary Allowances (1968.)

Table 42.
Foods for Special Dietary Use.

Nutrients	Infant Formulas		Special Products for Adults & Children	
	Soy Isolate Per qt 640 Kcal	Protein Hydrolysate Per qt 640 Kcal	MCT Dietary Per qt 960 Kcal	Tube Feeding Unit Dose 360 Kcal
Protein, g	24.0	21.0	33.6	22.0
Vitamin A, IU	2,000.0	2,000.0	4,000.0	1,670.0
Vitamin D, IU	400.0	400.0	400.0	133.0
Vitamin E, IU	10.0	10.0	15.0	10.0
Vitamin K, μg	100.0	100.0	(60)*.0	(80)*
Vitamin C, mg	52.0	52.0	80.0	35.0
Thiamine, mg	0.6	0.6	1.4	0.5
Riboflavin, mg	1.0	1.0	1.7	0.6
Niacin, mg	8.0	8.0	18.0	7.0
Vitamin B_6, mg	0.5	0.5	2.0	0.7
Vitamin B_{12}, μg	2.5	2.5	5.0	2.0
Folic acid, μg	50.0	50.0	100.0	33.0
Pantothenic acid, mg	3.0	3.0	10.0	4.0
Biotin, μg	30.0	30.0	50.0	(25)*
Choline, mg	85.0	85.0	135.0	(90)*
Inositol, mg	100.0	100.0	135.0	(90)*
Calcium, mg	900.0	900.0	1,000.0	500.0
Phosphorus, mg	650.0	625.0	800.0	450.0
Magnesium, mg	75.0	75.0	200.0	50.0
Iron, mg	12.0	12.0	18.0	6.0
Copper, mg	0.6	0.6	1.5	0.5
Zinc, mg	4.0	4.0	6.0	3.0
Manganese, mg	2.0	2.0	3.0	1.0
Iodine, μg	65.0	65.0	140.0	50.0
Sodium, mg	490.0	400.0	600.0	400.0
Potassium, mg	860.0	1,000.0	1,500.0	630.0
Chloride, mg	270.0	800.0	1,000.0	600.0

*( )— Not on label.

Table 43.
Meal Replacements (Instant Breakfasts).

Nutrient	Product A		Product B	
	Added	With Milk	Added	With Milk
Protein, g	X	17.50	X	18.80
Kcal	X	290.00	X	296.00
Vitamin A, IU	X	1,400.00	X	1,400.00
Vitamin D, IU		100.00		100.00
Vitamin E, IU			X	2.50
Vitamin C, mg	X	30.00	X	30.00
Thiamine, mg	X	0.30	X	0.35
Riboflavin, mg	X	0.57	X	0.72
Niacin, mg	X	2.70	x	2.70
Vitamin B_6, mg	X	0.37	X	0.4
Vitamin B_{12}, μg	X	1.17	X	1.20
Pantothenic acid, mg		(0.80)*	X	3.30
Folic acid, mg		(0.01)*		(0.01)*
Calcium, mg	X	410.00	X	560.00
Phosphorus, mg	X	370.00	X	460.00
Iron, mg	X	2.60	X	2.80
Copper, mg	X	0.19	X	0.50
Iodine, μg			X	25.00
Sodium, mg			X	283.00
Potassium, mg			X	721.00
Manganese, mg			X	0.50
Magnesium, mg		(30.00)*		(30.00)*
Zinc, mg		(0.70)*		(0.70)*

*( ) — Approximate levels supplied by milk.

Table 44.
Some Nutrients in Breakfast Cereals—Per Oz.

Nutrient	Cold				Hot	
	A	B	C	D	E	F
Protein, g	2.90	2.200	2.90	5.40	4.100	2.60
Thiamine, mg	0.10	0.120	1.00	1.00	0.020	
Riboflavin, mg	0.04	0.020	1.20	1.20	0.040	
Niacin, mg	2.20	0.600	10.00	10.00	0.300	
Vitamin B_6, mg	0.13	0.020		0.03	0.050	0.02
Vitamin B_{12} μg						
Folic acid, mg	0.03				0.010	0.04
Pantothenic acid, mg	0.30	0.060			0.500	0.16
Calcium, mg	10.00	1.000	70.00	90.00	12.000	300.00
Phosphorus, mg	145.00	9.000	65.00	150.00	110.000	300.00
Magnesium, mg	52.00	2.600	9.00	36.00	38.000	12.00
Iron, mg	4.50	0.400	13.00	2.00	1.200	25.00
Zinc, mg	1.20	0.040	0.20	0.80	1.100	0.60
Manganese, mg	1.20	0.040	0.30	1.00	1.100	0.30
Copper, mg	0.20	0.007	0.03	0.24	0.007	0.04

Table 45.
Essential Nutrients

Nutrient	US RDA	Nutrients per Meal	Approximate Amount/100 Kcal[a]
Protein, g	65.00	15-25	2.600
Vitamin A, IU	500.00	1250-2000	200.000
Vitamin D, IU	400.00	100-150	16.000
Vitamin E, IU	30.00	7.5-10	1.200
Vitamin C, mg	60.00	15-25	2.400
Thiamine, mg	1.50	0.4-0.6	0.060
Riboflavin, mg	1.70	0.5-0.7	0.070
Niacin, mg	20.00	5-8	0.800
Vitamin B_6, mg	2.00	0.5-0.8	0.080
Vitamin B_{12}, mcg	6.00	1.5-2.5	0.240
Folic acid, mg	0.40	0.1-0.15	0.016
Pantothenic acid, mg	10.00	2.5-4	0.400
Calcium, mg	1000.00	250-400	40.000
Phosphorus, mg	1000.00	250-400	40.000
Magnesium, mg	400.00	100-150	16.000
Iron, mg	18.00	4-6	0.700
Zinc, mg	15.00	4-6	0.600
Manganese, mg[b]	(4.00)	(1-1.5)	(0.160)
Copper, mg	2.00	0.5-0.8	0.080
Iodine, mcg	150.00	40-60	6.000

[a]Based on 2500 Kcal per day.
[b]No US RDA was set for manganese.

Q: This matter of riboflavin being optional in rice and calcium being optional in wheat products is confusing to the consumer. They think they are getting enriched bread and enriched rice but they don't know what they're getting. Some effort should be made to make a fortification standard.

A (Mr. Brooke): In Taiwan, riboflavin has been included in rice premix for the Nationalist Army since 1958. It has succeeded there because the rice is cooked in the Army mess, is thoroughly stirred before serving, and the yellow discoloration from the riboflavin is pretty well dissipated by stirring. That doesn't mean it will work in a civilian population.

Q: On what basis do you suggest fortifying cereals with protein for preschoolers? We are not familiar with information that indicates a great shortage of protein supply in our pre-school population.

A (Dr. Kline): This suggestion was merely for the inclusion of protein with other nutrients. We think, again, that the final addition of nutrients to foods of this kind needs to follow guidelines that are carefully considered, carefully established by a committee of the Food and Nutrition Board. Certainly, some breakfast cereals do contain protein, some contain added protein, and what was being protested in Chapter 13 was the practice of such a variability in composition that it is not in the interest of consumers of breakfast cereal.

Q: There has been some objection to the use of iodized salt due to allergies to iodine. Is that a serious situation?

A (Dr. Sebrell): At the time we were trying to get federal legislation to require iodized salt, this question came up. It was very thoroughly investigated by the Public Health Service at that time and an effort was made to track down individual cases to which people referred. In no case were they able to find a single individual who had any kind of reaction to iodized salt. There were persons to whom physicians had prescribed sodium or potassium iodine and, certainly, some people will get reactions to therapeutic doses of iodine. The quantity of sodium or potassium iodide in iodized salt, to the best of our knowledge, never has produced a single case of reaction.

A (Dr. Darby): Early concern about this problem arose from the observation that excessive amounts of iodine administered as a therapeutic agent can indeed produce, in certain instances, acneform lesions of the skin. When iodine was being added to salt initially, iodine and iodate were widely used as therapeutic agents in treatment of goiter and syphilis, and also in the treatment of hypertension and atherosclerosis. Some people got acneform eruptions because of the toxic levels they were receiving and the dermatologists became concerned about the addition of iodine to salt. But, no one has been able to verify any acceptable report of an allergy to iodine in the amounts that are used in iodized salt.

Q: If we get all the answers for zinc and for chromium requirements and want to add them to formulated foods, how do we go about it? At the present time, chromium and zinc are investigational new drugs; how do you get them classified as nutrients? Chromium is an investigational new drug at levels of 150 mg a day, which we consider to be required if it is given as chromium chloride. Zinc is an investigational new drug at levels of roughly 700 to 750 mg.

A (Dr. Sarett): We think you made a mistake by giving chromium as an investigative new drug at 150 mg. If that is supposed to represent the dietary requirement, you'll find it difficult to get the FDA to allow this as a regular food additive or as a nutrient. But, zinc, at 750 mg may be called a drug at those levels. We've used it at levels at 5, 6, and 7 mg in early Metrecal-type formulations. As long as animal and clinical experiments show that this is perfectly good, we don't see any problem in getting zinc classified as a food additive within safe limits or just using it at levels commensurate with what's in the diet.

Q: One reason why so few data are available on the nutrient content of processed foods with respect to degradation in processing, etc., is because restoration has been a word that has been avoided for years by nutrition policymakers. A few food manufacturers have restored nutrients in a quiet, non-advertising way and have had their wrists slapped despite the fact that they have not over promoted what they've done. Why is restoration apparently bad?

A (Dr. White): The Council and Board statements on addition of nutrients to foods do endorse the concept of restoration, among the recommended ways of maintaining or improving the nutritive quality of foods.

Q: We can't see that restoration has any particular merit about it. If we are designing a procedure for health purposes to maintain good nutrition, we think we should consider the quantity needed according to the Recommended Dietary Allowances or some other standard, rather than the amount that happened to be put into a seed or piece of meat. The goal of restoration may have no relevance to what we are trying to accomplish.

A (Dr. Johnson): We at FDA have had very few questions from processors specifically about restoration. On the one hand, does it make any difference if a food is restored to its natural value of 100th of the RDA for a given nutrient? Looking at it the other way, one could accept the argument that the natural composition of foods is important. We would agree—if we're using a four-food group approach to nutrition—then obviously, all the food, be it canned, frozen, or fresh that falls into a group, ought to be at some standardized level. There is a basic problem in that the so-called restoration level of one manufacturer is considerably higher than the restoration level of another. Both show data that suggest that the raw material was at one level or another. It does raise some practical problems if you call for actual labeling of content rather than a simple statement that says "restored to natural levels."

Q: Back in the 1930s and the 1940s, it was said that whole wheat flour was a much better product than white flour nutritionally, but consumers only wanted white flour. It was felt that nothing could be done to reverse the trend, so it was accepted and white flour was enriched. Now, we know today that white flour doesn't have anywhere near the B_6, the magnesium, and other nutrients of whole wheat flour. If white flour is such a widely used food in terms of breads, cakes, spaghetti and pizza, restoration of these other nutrients could help in getting them better disseminated to the public.

A (Dr. Sebrell): We never thought that whole wheat flour in bread was an exceptionally good product. We think there's a good bit of evidence that it isn't a very good product nutritionally, or not much better than white flour. The amount of riboflavin added in the enrichment formula for white flour exceeded that present in the whole wheat. We would take the viewpoint on vitamin B_6 now that we shouldn't restore to a level approximating whole wheat; but, rather, according to the human need for vitamin B_6, and what contribution to the total B_6 recommended allowance white flour products should properly make.

When we first talked about the enrichment of white flour, we used as a standard six slices of bread a day, which was the estimate of the bread consumption at that time. Two things have happened since then. The slices of bread are not as thick today as they were then and the average consumption is no longer six slices a day. So, perhaps we need to do something to the total enrichment formula to take these factors into account. This is just a possibility and we think similar arguments apply to potatoes and vitamin C. You would attack this problem on the basis of the human recommended allowance of vitamin C, what proportion of the vitamin C in the diet would be supplied by some average serving of potato or potato product. Vitamin C would be put into potatoes to make up the human need, not on the basis of what the potato happened to have in it before processing.

A (Dr. White): It is a valid point when we say restoration hasn't received adequate attention. The Council statement on improving nutritive qualities of foods endorses the concepts of enrichment, fortification, and restoration and illustrates what is meant by restoration with the following: "Restoration (of the lost nutrients) is acceptable when the product in question is nutritionally important, that is when the amount of the affected nutrients originally present provided at least 5 percent of the RDA in a serving." (*JAMA* 225: 1116-1118, Aug. 27, 1973.) When the Council presented its testimony at the Hearings on Foods for Special Dietary Use, the joint statement of the Council and the Board was presented as a concept for food design. We think it was the hope that the concept of restoration would be adopted as such.

Q: Should we not stress the importance of knowing how the food is used in the diet, and what segments of the population are consuming it, and in what quantities? We know that persons in higher income levels consume about 50 pounds of flour per capita annually. We know that people having incomes of less than $4,000 a year consume upwards of 400 pounds of flour a year per capita. Now, how are you going to fortify flour nutritionally? On the Navajo Reservation, as much as 80 percent of calories are derived from flour. There are data that indicate that probably the largest source of vitamin C in this country is not citrus fruits but potatoes. So we have to consider adding vitamin C to any processed potatoes, or some segments of the population are just going to be without it. A straight restoration is not the answer here.

Q: Chapter 10's nutrient-calorie ratios for fruits and vegetables showed enormous ratios of nutrients to calories, or calorie-RDA ratio. How do you handle the problem of a food with relatively low caloric density that, on a nutrient-calorie ratio, shows an enormously exaggerated contribution of nutrients?

A (Dr. Hansen): This system does tend to exaggerate the contribution to the diet of fruits and vegetables. One has to emphasize that Chapter 10's charts are incomplete because of the nutrients not listed, those that we don't know enough about. We would like to have taken the same kind of extrapolation and cut across the whole spectrum of nutrients that are known to be required biologically.

I would also like to comment on the previous question of restoration of nutrient content. I see no reason why we should not minimally restore a food as important as flour to the level of whole wheat. Because of the increase in consumption of foods which have low nutrient densities—snack foods—we must pay more attention to adding nutrients, even above the level of restoration.

A (Dr. Sarett): We think we get ourselves tangled in the semantics of the terms enrichment, fortification, supplementation, restoration. What we must really be concerned with is the usefulness of these additions and these foods to the ultimate consumer. This comes right back to Chapter 8's position with respect to the Recommended Dietary Allowances and to the needs of the individual who consumes the foods that we're trying to prepare; it little matters what we call the process of addition so long as the total diet meets the nutrient requirements for optimum health.

Q: We are concerned as to what the nutrient-calorie ratio on a basis of 100 calories or 1,000 calories means to the ultimate consumer? Chapter 10's charts were excellent because they are simple and understandable, but they should be placed on a per-serving basis so the consumer can know what he's eating. For example, dill pickles contain only 11 calories per 100 grams, but have a very high protein, calcium, and phosphorus content.

A (Dr. Hansen): The concept can be misused. This may be how Dr. Linus Pauling arrived at his estimate of the human vitamin C requirement—he may have looked at vitamin C in proportion to calories in foods containing high levels of vitamin C. You should take into consideration average serving, but when you are dealing with the consumer you must speak about food quality in very general terms. I don't think you can deal with a consumer in micrograms, milligrams, grams, and international units. But, if you talk to a consumer about broad qualitative values, you can usefully describe a food. Using the nutrient-calorie, whether you use a 100 or a 1,000 calorie base, the ratio comes out the same.

If the nutrient-calorie ratio is 10 and the food constitutes 10 percent of calories, the food provides the day's requirement for that nutrient. If the nutrient-calorie ratio is 30, as it is for ascorbic acid in orange juice, you can take one-thirtieth of your calories from oranges and get a day's requirement of ascorbic acid.

Q: How are you going to tell the consumer this story? This is the thing that's bothering me.

A (Dr. Hansen): We are not so sure you can't put a picture on a label which is certainly equivalent to, if not more informative than, a number. The number is the thing that bothers us. We don't know what a consumer gets when he sees MDR and

RDA with milligrams, grams, and micrograms. That kind of labeling is not very useful. If you can give the consumer the kind of labeling that provides a broad approximate notion of a quality judgment about a food, then it begins to make sense.

Q: In light of the startling figures in eating habits of the public and our difficulty in keeping up with them, cannot vitamin and mineral supplements that are necessary for good nutrition be considered in the form of a pill?

A (Dr. Johnson):   If we could get people to take a pill, we probably could get them to do some other things for good nutrition. But we seriously question that we can get them to take a pill on a regular basis. We prefer the concept of spreading nutrients throughout the diet so that as everybody meets their caloric needs, they can, in fact, get their total nutrient intake. This means using some rational approach to fortification, for there is a physical limit as to what most people can eat. If one needs to take 10 pounds of food in order to get a perfect diet, it's just not going to happen. It's very difficult, in our estimation, to visualize a successful pill but it may be not as difficult to talk about a formulation of food products in which you can get, let's say, 10 to 15 percent of the requirement in every 8 to 10 percent of the calories. We are not sure we're at the point yet where we can do that.

Q: We are proponents of enrichment procedures, but we want to point out that there are opponents to nutrient supplementation. We attended the recent Iron Workshop at which an eminent hemotologist told us that there are certain dangers for a small proportion of our people in iron fortification. Can you nutritionists actually show that a slight anemia, of 12 grams or so of hemoglobin, is bad for you? A slight anemia may increase your physical efficiency, your heart output. We must realize that in nutrient fortification we are dealing with a balance of a potential beneficial effect versus possible detrimental effects to some. To justify going ahead we need a clearcut recognition of the marginal deficiency states, and what they do to a person.

A (Dr. Combs):   We would also like to identify ourself as proponents of enrichment, fortification, or restoration. But we'd also like to discuss the concept of the pill before it is dismissed entirely. We need to approach our nutritional problems from a preventive medicine point of view. We can't wait for clear evidence of need to justify the addition or restoration of nutrients such as pyridoxine where one could prevent considerable malnutrition in our population. Other nutrients that we would like to see added would include zinc at a very nominal, safe, preventive level. If this supplementation can be done without too much delay through our food supply, that obviously is the direction to go. Perhaps, before one goes to a pill, a drink might be used to provide supplemental preventive levels of various nutrients.

We do have problems of sub-optimal intake, we have health risks that could be removed at a much lower cost than treatment of a chronic disease arising from a long-term, sub-optimal intake. We may have to divide nutrients into those that can be added at safe levels, are inexpensive, and can be controlled, and those which must be handled on an individual basis. A chewable pill would be the cheapest form for the latter, as a short-term measure. The long-term objective clearly should be supplementation through our food supply, coupled with appropriate nutrition education measures.

A (Dr. Babayan):   We'd like to make two other comments. First, you could have the most nutritionally balanced food and be unable to get public acceptance. Second, manufacturers of fats and oils have put forth great effort and expense to eliminate metallic contaminants that affect stability, color, and flavor. If iron, copper, zinc, and other trace elements were put into fats and oils, they might be quite satisfactory nutritionally but totally incompatible with good food practices and good food stability.

Q: In regard to pills, let's not overlook the fact that about one-third of prescriptions are not filled and about one-third of those that are filled are not taken. We are not sure that the pill is going to answer the nutrition problem.

A (Dr. Brin):   We should consider cost, too. It would be quite difficult to make a pill-type supplement to cost less than a penny a day, but foods such as flour can be fortified with a hundredth of a cent for the same amount of nutrient. It will require a judicious choice of food varieties to convey the necessary nutrients.

We have heard the argument that

vitamin A should be added only to milk or skim milk but not to anything else because we can train people to consume yellow vegetables. Well, anyone who has ever tried to do this knows how futile it is. The Puerto Rico Nutrition Committee has tried for 25 years to get people to buy all varieties of foods and to consume larger amounts of yellow vegetables. Now they're actually considering adding vitamin A to flour in order to satisfy those lacks which were revealed by their National Nutrition Survey.

Chapter 8 hit right on the spot what we consider a generation gap in nutrition and nutritional attitudes in this area of enrichment. It seems to us that all of the arguments which have been raised against adding nutrients to foods which are generally distributed were faced by our previous generation, 25 and 30 years ago, and they solved it. Chapter 8 said that we must solve public health problems at minimal cost without making a person buy anything new, without changing dietary habits, without changing flavor or texture of food.

A (Dr. Sebrell):   The additional nutrients we have to deal with today are complicating the problem of supplementation rather than making it easier. We'd like to call attention to the one thing in the RDA that some may not have recognized. The recommendation for iron for the adult woman is 18 mg. The number of calories recommended for that woman is 2,000. The amount of iron in the usual American diet, we are told by the iron experts, is around 6 mg per thousand calories, so we have consciously made a recommendation that can't be met by natural foods. We recognized the dilemma when we created it, but it was the proper answer.

There's still a place, in my opinion, for a maintenance nutritive pill for some people. Chapter 16's data indicated that a large number of people in this country who are concerned about obesity are restricting their calories. It's quite obvious, the more you restrict your calories, the more difficult it is to meet your nutritional requirements. Many people are on very strict therapeutic or reducing diets in which a vitamin pill in a maintenance dose may be a valuable thing. But the pill is not the answer for the general population—it is for individual cases.

reports of the task forces

PART FOUR

Charge To identify, by order of priority, those areas of vitamin and mineral human nutrition (including definition of requirements) where new or additional research is indicated as being essential to the establishment of any comprehensive program of effective, definitive guidelines for micronutrients in processed foods, and to suggest avenues by which the research could be conducted.

PREFACE

The task force found a fundamental disagreement with its charge. The charge suggests that research is a prior necessity to establishment of effective programs or guidelines. While it is true that a great deal of research needs to be done, sufficient indications are at hand to allow the establishment of tentative guidelines and programs. The committee thus suggests that the research indicated below has extremely high priority but need not be completed prior to the establishment of programs where sufficient knowledge is available to constitute a basis for action.

Increasing rhetoric with reference to nutrition seems to be inversely proportional to the expenditure of funds for definitive programs, training, or very badly needed research. If this trend continues, we shall end up by having no ongoing nutrition programs and numerous public debates based upon data which are outdated. The result of this inverse relationship between stated concern and allocated funds is shortsighted in the extreme.

Applied research in human nutrition seems to have decreased in priority in recent years. The major support for nutritional research has come from the National Institutes of Health where, unfortunately, the trend in funding of research in nutrition seems to be toward basic biochemistry and mechanistic studies rather than applied human nutrition. As a consequence, we are making very slow progress in developing physiological and biochemical data that may be utilized to better define the recommended dietary allowances for man of known required nutrients, and equivalently slow progress in determining whether or not additional micronutrients are needed for complete human nutrition. Unless this trend is reversed, we shall not be able to establish

---

chapter 18
needs for
future research

Report of Task Force I

---

definitive programs to optimize human nutritional status.

Perhaps a better identification of human nutrition should be developed in either the Department of Health, Education, and Welfare or the Department of Agriculture. It is extremely important to develop a program in foods and nutrition which is not dependent upon groups such as the National Institute of Arthritis and Metabolic Diseases. Foods and nutrition should be integrated to allow development of the applied aspects of physiology and biochemistry.

RECOMMENDATIONS

Priority I In order to develop a better definition of the RDA, a substantial effort is needed to bring together the available data on vitamin and mineral requirements of man and the availability of these vitamin and minerals from foods. It will be necessary to study the consequences of marginal nutritional states and develop criteria of accuracy by:

1. Monitoring programs of supplementation which are already in existence.

2. Epidemiological surveys of populations which are in a marginal nutrition state.

3. Investigation of the biochemical and/or physiological parameters concerned with the low levels of nutritional intake.

In this area we do not have sufficient data on the minerals magnesium, iron, manganese, calcium-zinc, and the calcium-phosphate ratio. We are aware that this list is not exhaustive, but in order of priority these items seem to the task force to be the most important. In vitamins equivalent work is needed, again in order of priority, on folate, thiamin, A, B_6, B_{12}, E, K, pantothenic acid, and C.

Priority II In order to develop programs of enrichment or to understand where they may be needed, it is important to determine the availability of vitamins and minerals to man from the food which he eats. Thus, we should:

1. Study the bio-availability of vitamins and minerals of foods that are consumed by the population. Data are needed on bio-availability in foods as eaten, particularly in the new engineered foods.

2. Determine the physical factors affecting the bio-availability of micronutrients in foods.

3. Determine the bio-availability of vitamins and minerals in mixed diets, as there is evidence that mixtures of food normally may alter the availability of the micronutrients.

4. Determine the effects of storage conditions and unit processing operations on the bio-availability of vitamins and minerals. This is particularly important where modified-unit processing operations are developed. Priority should be given to foods which are the major sources of iron, vitamin B_6, zinc, folic acid, vitamin E, pro-vitamin A, calcium-zinc, magnesium, and calcium/phosphate.

Priority III Field studies geared to the evaluation of enrichment programs on populations are urgently required. Both biochemical and physiological studies are needed in this area. Equivalently important is a better definition of dietary habits

and food consumption patterns. These should be analyzed with respect to age, sex, ethnic background, and socioeconomic status.

Priority IV Since a very high percentage of our total food intake in the United States is processed, the industrial problems of enrichment are of particular importance. Enrichment may interfere with the success of a product in the marketplace by contributing to off-color or off-flavor. Thus, the technical problems of fortification with vitamins and minerals must be studied in relation to flavor, color, and texture. Although some of this work is done in industry, broader studies can be conducted in universities or institutes with expertise in food science and technology.

Priority V Substantial work is needed on the development of analytical methods to determine bio-availability. Although this may be inherent in Priority II research, it was thought worthwhile to explicitly break this particular problem out:

1. Especially pressing is the need for the development of chemical analytical methods which show a good correlation between the bio-chemical availability of vitamins and minerals and the chemical methods.

2. Rapid methods for analysis of vitamins and minerals are particularly needed for use in epidemiological work.

3. More sensitive methods for the micronutrients are needed. In many cases, methods now available are not sufficiently sensitive to indicate the quantitative bio-availability of minor constituents of foods, and yet these may be at the physiologically active levels.

Priority VI Projects concerned with identifying the remaining essential elements in human nutrition should be encouraged and supported.

Priority VII Investigations of food as eaten should be conducted to determine the role environmental contamination may play upon the total mineral content of the food. Particular attention should be paid to selenium, copper, cadmium, mercury, lead, arsenic, and cobalt.

Priority VIII Substantial research is needed to establish safety margins for the minor essential elements, particularly selenium, copper, cadmium, lead, arsenic, cobalt, and magnesium. There are substantial differences between the compounds of these elements with respect to toxicity. It is imperative, therefore, that the compounds normally found in food be identified as compounds rather than as their parent minerals and the toxicity of these determined.

METHODS OF ACHIEVING THE NEEDED RESEARCH GOALS

The main impediment to achieving research goals is the lack of financial resources to support the work. Priority appears to have been given to biochemical aspects of nutrition in the United States, with very little attention being given to the physiological and applied nutritional areas.

We, therefore, recommend that an agency be created within one of the established departments of government (perhaps in the Department of Agriculture) which would intramurally and extramurally support work in nutrition, food science, and food technology. It should develop a major focus on problems directly related to applied nutrition. While this proposal has been reiterated many times, it is not likely to be implemented until the combined interests of industry, government, and academic institutions are used to promote the establishment of such an agency. Even with combined effort, an identifiable institute or agency will take substantial time to come to fruition. Obviously, the implementation of such an agency also would require the development of a national nutritional policy which, so far, we have not been able to achieve.

The problems of nutrition are extremely pressing and cannot wait for the necessary bureaucratic reorganization to initiate substantial programs. We recommend that an external agency concerned with solving some of the more important priority problems in nutrition take over the interim chore. The AMA Council on Foods and Nutrition, together with its industry liaison panel, might assume the task. This group, with its direct concern for the applied aspects of nutrition, could act as a focus for establishing an extramural program in research. A group of this type could solicit funds from industry, foundations, and governmental agencies to promote research in the areas pointed out above.

In general, laboratories such as those being established by the Department of Agriculture could achieve the research aims noted by the task force. The obvious other source of research expertise would be the universities with departments of nutrition and food technology which could, given the funds, attach the problems of high priority in the nutrition program.

Charge To identify factors which must be taken into account in the development of guidelines for the preservation or addition of vitamins and minerals in processed foods. These would be based upon:

1. Consumer understanding and acceptance.
2. Manufacturing technology.
3. Distribution practices.
4. Economic considerations.

The panel discussed each of these four areas and reached the following conclusions.

CONSUMER UNDERSTANDING AND ACCEPTANCE

A potential problem of sub-optimal micronutrient intake is developing due to a low degree of understanding of nutrition and nutrient contribution of foods by consumers, changes in types of foods being produced, and changes in consumer consumption patterns. This provides the opportunity for a program of adding micronutrients to foods on the basis of agreed-to scientific evidence, even though all of the facts may not be available at this time.

As a part of any program for the addition of micronutrients to foods, we recommend that a suitable body of scientific evidence be assembled as a supporting basis. We also recommend that a fortification program be based upon opportunities provided through basic food consumption patterns so fortification can be selectively provided without the need for extensive educational effort to change food consumption habits. Selective fortification of foods could be based on their contributions to recommend dietary allowances, and simplified label statements referring to this contribution to RDAs could serve as the communication to the consumer.

MANUFACTURING TECHNOLOGY

We recommend that a fortification program take into consideration the impact that micronutrient additions might have on flavor, texture, color, and stability—all of which are food characteristics important to consumer acceptance of food. All foods do not provide equal fortification opportunities due to the aforementioned characteristics. There-

chapter 19
fortification of processed foods

Report of Task Force II

fore, it will be important to identify appropriate groups of foods which can provide opportunity for selective fortification.

Such factors as consumption patterns (either across the population or by population groups with special dietary problems), the ease of addition of nutrients to products, the availability and stability of micronutrients in foods, and the influence of the processing and storage of products on the stability and availability of nutrients are important criteria to be considered in selecting appropriate foods for fortification. Current technology will permit selective fortification and we recommend that special problem areas be identified so the technology can be developed where necessary to permit micronutrient additions where important. Some technological problems may take time to solve. This should be recognized in fortification recommendations and procedures should be outlined whereby the necessary development work can be accomplished.

DISTRIBUTION PRACTICES

Many of the same considerations discussed under Manufacturing Technology apply to distribution practices. The length of time that a food is in the distribution system is an important consideration. Therefore, nutrient and organoleptic stability must be related to time. Current food manufacturing and distribution practices frequently require long distribution times for greatest economic value to the consumer. Therefore, an appropriate balance of nutrient and organoleptic stability relative to the needs of the distribution system has to be considered.

While the US as a whole is a very large country, opportunities for regional policies should be considered where certain regional and/or ethnic food habits will provide an opportunity to reach specific population groups. The panel notes the tendency for food fortification regulations to be developed by the individual states and feels that it is in the best interests of both the consumer and the processor to have a uniform application of food fortification regulations. This is important when considering the impact of regulations on consumer education effort as well as the uniform application of nutritional principles in the fortification of foods.

ECONOMIC CONSIDERATIONS

It is important to look at the cost of micronutrient fortification. When evaluating costs of fortification, we recommend that the following be considered:

1. Micronutrient costs themselves generally are not prohibitive.

2. Analytical costs generated by micronutrient fortification could be a potential problem for the small food processor.

3. Consideration should be given to developing simplified methods of quality control to minimize analytical costs.

4. The costs of equipment, processing, and formulation changes should be considered when evaluating the total potential economic impact of fortification.

5. When considering the fortification of snack foods, protein content should be a key factor in deciding on fortification opportunities. The cost for this nutrient would have already been built into the product and the added micronutrient costs would then be quite small.

There is no economic reason why government and industry, working together, cannot make available a nutritionally adequate diet for all consumers.

Charge In the context of knowledge regarding US meal patterns and food preferences, to develop recommendations and priorities for educational labeling or other approaches to assist consumer population groups to meet micronutritional needs through appropriate buying patterns and food choices.

To a degree, the charge to the task force could have been better carried out after the reports of the other four task forces were received. Not having that luxury, work was undertaken by way of visualized concepts that should lend themselves to appropriate interpretation in the light of the recommendations of the other groups.

The task force recognized that it was working in large measure without having available specific knowledge of meaningful correlations between nutritional status and actual eating habits. We do not know why people who are found by clinical or biochemical evaluations to be "ill-nourished" arrived at that state. Information relating dietary habits to nutritional status is not yet available from the national nutrition surveys. Nor do we know to what degree marginal states of nutritional status affect performance.

RECOMMENDATION I

Consumer understanding of such terms as "iodized," "enriched," "restored," "fortified," etc., is fuzzy. For example, does the consumer know the difference between iodized salt and plain salt? Is it really meaningful to today's consumers to depend upon label identification alone to encourage the purchase of the iodized product? The task force thinks not. Therefore, we recommend that:

> For effective nutrition education programs the consumer must have a clear understanding of the terminology used; for example, iodized, enriched, fortified, etc., need clear definitions and consistent usage.

RECOMMENDATION II

The examples cited above exemplify the need for continuing programs of education to encourage the use of a variety of foods, some of which may include fortified and enriched staple foods (ie, foods chosen as vehicles for nutrients

chapter 20
nutrition education of consumers

Report of Task Force III

in "short supply in the general dietary"). We recommend:

> In the field of preventive or corrective nutrition no single effort will achieve desired results. Nutrition education programs must be instituted concomitantly with these efforts.

RECOMMENDATION III

Corrective or specific preventive measures which involve foods and nutrition must be based upon solid information and be designed to accomplish the greatest benefit. Oftentimes proposals are put into practice without adequate attention to their actual impact on the consumer. Other times significant programs are undertaken without appropriate baseline studies which would permit later investigation of the effectiveness of the program. Under these circumstances it is difficult for the educator to assist in the support of the programs and to evaluate effectiveness of current approach. The task force recommends:

> When feasible, consumer research should be done to evaluate proposed educational and regulatory practices.

RECOMMENDATION IV

The development of meaningful programs in public education in nutrition requires agreement on the important concepts. While some of the concepts are presently based on informed judgement, most are founded on solid fact. Educators and others who do not possess a knowledge of the science of nutrition can teach and develop useful materials if they are provided with a compendium of basic concepts. We recommend:

> A compendium of basic concepts be developed and given wide distribution to key thought leaders. These concepts should be disseminated locally but have national backstopping.

RECOMMENDATION V

The fifth recommendation deals with both a clearer approach to professional education and a plea for some efforts to inform the professional person to teach nutrition to others:

> A clearer approach to professional education in the area of nutrition should be developed. The medical and paramedical community should have a better understanding of the relationship of nutrition to health, and instructors should be taught how to teach nutrition in elementary education.

RECOMMENDATION VI

Nutrition, when taught, is often done so in a helter-skelter method. If the teaching of nutrition in public education is to succeed, the whole subject should be laid out in a planned, sequential manner. Every effort should be made to assure that the material be laid out sequentially so as to prevent boredom from excessive repetition.

The School Health Education Study-Curriculum Development Project is a leading group making great progress in developing a conceptual approach to curriculum design in health education for kindergarten through twelfth grade. Their efforts regarding nutrition education should be evaluated and supported when found to be practical. The task force recommends:

> Every effort should be made to develop and implement a sequential approach to teaching basic nutritional concepts, at all levels of education.

RECOMMENDATION VII

The task force devoted considerable time to discussing various "schemes" for teaching the combination of foods into nutritious meals. It was obvious that no single concept could embrace all of the possible food combinations available today. For the most part, the task force members agreed that effective variations of the present basic food groupings could be developed. A difficulty with that concept is identification of a main course item which does not happen to be meat. Another difficulty is that in reality the four groups are seven distinct groups—dairy foods; meat and alternatives; cereals; fruits and vegetables; special emphasis given daily to fruits and vegetables containing vitamin C; special emphasis given every other day to fruits and vegetables containing vitamin A; and all other foods not meeting the above criteria.

Within the task force recommendation is a sub-recommendation suggesting special consideration of the "fruit-vegetable" group. That many people "avoid" vegetables suggests that insufficient emphasis is given to their educational promotion. The task force recommends:

> Education approaches that stress a wide variety of foods in the usual meal patterns are vital. There is need, however, for better information on how to blend foods into nourishing meals. Identification of a meal main dish (entree), based upon protein, calories, and appropriate nutrient density would help. The completion of meals then might be in keeping with the concept of food groups. One way to accomplish this requires better definitions of the contributions of nutrients accept-

ed from the fruit-vegetable group. Ultimately this may mean natural and artificial modifications of their nutrient contribution to avoid the present divisions into three subgroups.

RECOMMENDATION VIII

While the task force was not prepared to suggest alternative possibilities, it was concerned that food labels with quantities of nutritional data would be uninformative to the average consumer. The task force recommends:

> Doubts about the usefulness of nutritional labeling (nutrient data) to the consumer were expressed. Other concepts should be explored before locking into a complex label system.

RECOMMENDATION IX

The task force was informed of a new project being developed by the Society of Nutrition Education. It was understood that an archive of available educational materials is being developed. Since there is considerable duplication of effort in the production of educational materials, such an archive should prove to be most valuable. The task force recommends:

The task force is aware of the development of a library of nutrition education material instituted by the Society of Nutrition Education and encourages the Society to consider the recommendations made by this task force.

RECOMMENDATION X

Although mass information programs can reach the greatest number of people, provision should be made for teaching in small group assemblies. In the long run, no educational effort is superior to person-to-person contact. There are many examples of programs that feature one-to-one teaching experience. In-store consultants can greatly assist the harrassed shopper. The task force recommends:

> The effectiveness of direct contact with the consumer through nutrition aides and printed material is recognized and the task force encourages the expansion of the programs. The assistance of industry, consumers, and other appropriate groups or agencies is encouraged.

Charge To identify the function of regulatory control in establishing and maintaining effective nutritional quality guidelines, and to identify the basic considerations for making nutrition guidelines under regulatory control complementary to voluntary industry programs, and to nutrition education of the consumer.

Because of their collateral relationship to this task force's area of principal concern, some areas of primary concern to other task force groups have been touched upon. Our recommendations are:

1. The basic objective of the regulatory function is to avoid hazards and to prevent fraud and deception. However, regulations must be such as to permit a variety of foods and labeling that will allow consumers a freedom of choice in selecting foods to make up adequate and acceptable diets. The panel feels that there cannot be accomplished by regulatory means that which must be done by educational effort.

2. Regulations must recognize the need for the addition of nutrients to foods that are represented for general use.

3. Nutritional guidelines should be established for certain food categories. These guidelines should be based upon the best scientific information currently available. Such guidelines must be regarded as provisional and subjected to periodic review.

4. Such guidelines should provide criteria for informative nutritional labeling, using a system that will be applicable to all classes of foods.

5. The marketing of a food that is represented as meeting the established guidelines must be permitted on a voluntary basis; but, if it is so represented, it must be in conformance with the nutrient composition and labeling requirements established in the applicable guidelines.

6. The haphazard, indiscriminate, or irrational addition of nutrients to foods should be prohibited.

7. A system of nutritional evaluation based upon the relationship of nutrient content to caloric value of foods should be considered.

8. It should be recognized that the nutrient content of a food is equivalent whether it is achieved naturally or by the addition of nutrients.

9. The panel strongly recommends the development of guidelines and regulations

chapter 21
regulatory aspects of nutrition guidelines

Report of Task Force IV

through cooperative efforts based upon objective and scientific consideration, with avoidance wherever possible of adversary proceedings. Further, the panel abhors the intrusion of emotionally charged pressures from political, consumer, academic, or industry sources into the due process of administrative or regulatory proceedings related to nutrition matters.

Charge To state the rationale for development of guidelines for vitamin and mineral quality of processed foods, and to identify criteria upon which to base guidelines.

The time available did not permit consideration in depth of all points identified for attainment of unanimous positions relative to some. This summary indicates the points upon which there was broad agreement and, in some instances, unresolved differences.

The task force agreed that the Minimum Daily Requirements (MDR) have outlived their usefulness and that the Recommended Dietary Allowances (RDA) of the Food and Nutrition Board provide the best basis for establishing guidelines for nutrient standards. It further was agreed that criteria of nutrient composition of processed foods should not be limited to those nutrients for which allowances are depicted in the table of the RDA report. Consideration must be given to other nutrients discussed in the Recommended Dietary Allowances and recognized as required by man but for which quantitative information is insufficient to permit inclusion in the RDA table.

A few of the latter category of nutrients may be placed in a more restricted category. These are the nutrients such as selenium or fluorine for which the margin between the allowance and the toxic level may be very narrow.

The task force agreed that in a revision of the Recommended Dietary Allowances the discussion of less-well-understood nutrients should be as extensive and complete in order to provide better guidelines for these and other purposes in the future. The task force agreed that the exact criterion or guideline to be applied in a specific instance depends upon the category of food, its use pattern, and the objective of the fortification. For its discussion of these concepts, the panel found it necessary to try to define types of food somewhat more clearly and tentatively used the following definitions:

1. Conventional foods are in general those items that are identified as among the basic four or basic seven groupings as indicated by the US Department of Agriculture.

2. Formulated foods consist of two or more ingredients other than seasonings that are processed or blended together.

chapter 22
rationale and criteria for micronutritional improvement of foods

Report of Task Force V

3. Fabricated food is an analogue of traditional food which involves duplication of the structure and system of the foodstuff.

The panel agreed that the nutritional improvement or protection of conventional foods should be permitted and that a demonstration of a deficit or nutrient deficiency within the population should not be a prerequisite for permissive improvement of conventional foodstuffs. The task force holds, however, that promiscuous or excessive fortification is to be avoided. The task force opposes power races in nutrient improvement of foodstuffs as not in the best interest of the public. It was agreed that:

1. Useful and suitable guidelines for designing fabricated foods and for total meal replacements (certain special pur-

pose foods) are provided by the ratio of the Recommended Dietary Allowance for a nutrient/1,000 calories (or 100 calories). This may be referred to as the RDA-calorie ratio or nutrient density.

2. Another useful criterion or guideline for improvement or protection of conventional processed foods is the nutrient composition of the traditional unprocessed foodstuff or commodity. The term, "nutrient protection," of a conventional processed food refers to prevention of loss of nutritional quality such as might result from a changed or new processing technique. The use of known nutrient composition of traditional and processed foodstuffs or commodities as a guide for nutritional improvement has long been used and termed "restoration."

3. The criterion for design of food analogues should be the average nutrient composition of the class of conventional foodstuffs displaced or simulated. In applying these guidelines, it was agreed that the nutrients present in conventional foods in "dietary inconsequential amounts" need not be included in the fabricated analogue. It is suggested that an "inconsequential amount" might be defined in terms of quantity less than a given percentage of the Recommended Dietary Allowance of the nutrient per 1,000 (or 100) calories.

Some members of the panel felt that this limit might be set at less than 25 percent of the RDA-calorie ratio; others thought this level should be lower (ie, 20 percent or so). The list of nutrients that it was agreed should be considered currently for such usage is: protein, vitamin A, vitamin D, vitamin E, vitamin C, thiamine, riboflavin, niacin, vitamin B_6, vitamin B_{12}, folic acid, pantothenic acid, calcium, phosphorus, magnesium, iron, zinc, manganese, copper, and iodine. The quantity and composition of the fat would need to be considered.

To illustrate the application of these concepts, the panel indicated that snacks and supplementary food should be permitted to be fortified on the basis of the RDA-nutrient calorie ratio for those nutrients identified above. On the other hand, foods intended as total dietary replacements to be used exclusively over long periods of time are special dietary foods, and should be formulated to include an extended list of nutrients in order to meet all potential nutrient requirements of man.

The panel repeatedly emphasized that nutrients added to products in any form of fortification, but particularly into food analogues, should have a biological availability such that the quantity absorbed is equivalent to that amount available from the conventional food that may be displaced.

4. The panel recognized the continued importance of the concept of enrichment designed to eradicate a specific obvious deficit of a particular nutrient (or nutrients). The vehicle (or vehicles) selected for such enrichment must be consumed in appropriate amounts by the target segment of the population and the level of nutrient added determined by the usual daily intake of the vehicle by the target group.

The level must be sufficient to introduce preventive quantities into the diet but not high enough to produce either imbalance or toxicity. Such enrichment should be based upon acceptable evidence of need and must be economically feasible. The task force felt that a rational approach to fortification can be based on the following concepts:

a. A diet based on the basic four food groups will, if reasonably well balanced among the four, provide a fully adequate supply of micronutrients that will correspond (with the exception of iron) to the Recommended Dietary Allowances

b. A processed food fabricated from purified (or isolated) carbohydrates, protein, or fat may be deficient in vitamin or mineral micronutrients and, unless appropriately fortified, could introduce an imbalance in the dietary relative to the Recommended Dietary Allowances.

c. The Recommended Dietary Allowance for micronutrients can be expressed in terms of averages of the nutrients per 1,000 calories or per unit of calories, and these averages used as guidelines for the fortification of processed foods, such as previously discussed. Attention was given to the question of whether all group averages of Recommended Dietary Allowances should be used or maximum nutrient density for the particular category of consumer as the basis for guidelines. The panel was not completely unanimous in its position as to the actual basis of such guidelines, but it was noted that there is little practical difference in the standards derived by the two methods.

d. A processed food fortified by this means when combined with a dietary otherwise based on the basic four food groups will not contribute to an imbalance and therefore could justifiably be described as being *balanced with respect to the RDA*.

e. Processed foods fortified with micronutrients will introduce neither an excess nor deficiency in the dietary if levels of fortification are within 20 percent of the averages indicated above. Such fortified foods can be accurately described for all practical purposes by the simple label statement that they have been "balanced by fortification to conform to the current RDA."

f. There was considerable opinion that since individual foods or food groups of the basic four are not completely balanced with respect to all of the nutrients included in the current RDA, then it would seem reasonable to recognize some comparable latitude for fabricated foods. Accordingly, some members of the panel felt that fabricated foods could be "balanced with respect to B-vitamins," "balanced with respect to minerals," "balanced with respect to fat soluble vitamins (except D)," "balanced with respect to vitamin C," or "balanced with respect to protein." Other members of the panel were of the opinion that such balance in but a portion of nutrients would be confusing to the consumer and should not be practiced.

The panel recognized that other considerations may pertain in designing specific foods. For example, a nutritionally complete, quick breakfast could be designed using the guidelines of RDA nutrient-calorie ratio, except that it may contain more than that amount of vitamin C, in as much as the "traditional" breakfast contains a high proportion of the daily vitamin C intake for many persons. Similarly, low-calorie meal replacers may be designed on the RDA nutrient-calorie ratio basis, but properly utilize as a calorie basis a "usual" calorie intake rather than the low-calorie quantity represented by the meal replacer.

participants

APPENDIX

participants in the symposium on nutrients in processed food

Philip D. Aines, PhD, The Proctor & Gamble Co., Winton Hill Technical Center, Cincinnati, Ohio; Task Forces' Chairman

Bernard Alexander, Beech-Nut, Inc., New York, N.Y.

Ray H. Anderson, PhD, General Mills Technical Center, Minneapolis, Minn.

Lawrence Atkin, PhD, Arthur D. Little, Inc., Jensen Beach, Fla

John C. Ayres, PhD, Food Science Department, University of Georgia, Athens, Ga.

Vigen K. Babayan, PhD, Stokely-Van Camp, Inc., Indianapolis, Ind.

Richard H. Barnes, PhD, Dean, Graduate School of Nutrition, Cornell University, Ithaca, N.Y.

Howard E. Bauman, PhD, Vice-President for Science and Technology, Pillsbury Co., Minneapolis, Minn.

Evan F. Binkerd, Armour & Co., Oakbrook, Ill.

John J. Birdsall, PhD, WARF Institute, Inc., Madison, Wis.

Benjamin Borenstein, PhD, Roche Chemical, Hoffman-La Roche, Inc., Newark, N.J.

James L. Breeling, Director, Department of Scientific Assembly, American Medical Association, Chicago, Ill.

Myron Brin, PhD, Hoffman-La Roche, Inc., Newark, N.J.

William A. Brittin, Stange Co., Chicago, Ill.

Clinton L. Brooke, Consultant, United States Agency for International Development, Raleigh, N.C.

Ann L. Burroughs, PhD, Del Monte Corp., Walnut Creek, Calif.

J. A. Campbell, PhD, Assistant Director, General Nutrition, Canadian Food and Drug Directorate, Ottawa, Ontario, Canada

C. O. Chichester, PhD, Professor of Food Science, Department of Food and Chemistry, University of Rhode Island, Kingston, R.I.

Gerald F. Combs, PhD, Coordinator for Nutrition and Food Safety, Office of the Secretary of Agriculture, Washington, D.C.

Gary E. Costley, PhD, Director of Nutrition, Kellogg Co., Battle Creek, Mich.

Robert H. Cotton, PhD, ITT Continental Baking Co., Inc., Rye, N.Y.

William J. Darby, MD, PhD, President, Nutrition Foundation, Inc., New York, N.Y.; and Professor of Biochemistry and Nutrition, Vanderbilt University School of Medicine, Nashville, Tenn.

Tom Davis, Merck and Co., Rahway, N.J.

Philip H. Derse, MS, President, WARF Institute, Inc., Madison, Wis.

Norman W. Desrosier, PhD, New Products Services, Inc., Ho Ho Kus, N.J.

William E. Dickinson, President, Salt Institute, Alexandria, Va.

Kenneth G. Dykstra, PhD, General Foods Corp., White Plains, N.Y.

Donald F. Emery, PhD, Technical Director— Quality, General Mills, Inc., Minneapolis, Minn.

Richard P. Farrow, Assistant Director, Washington Research Laboratory, National Canners Association, Washington, D.C.

Lloyd J. Filer, Jr., MD PhD, Professor, Department of Pediatrics, University of Iowa Hospitals and Clinics, Iowa City, Iowa

Ray C. Frodey, Gerber Products Co., Fremont, Mich.

Grace A. Goldsmith, MD, Dean, Tulane University School of Public Health and Tropical Medicine, New Orleans, La.; and Chairman, American Medical Association's Council on Foods and Nutrition

Willis A. Gortner, PhD, Human Nutrition Research Division, Agricultural Research Service, United States Department of Agriculture, Beltsvelle, Md.

D.M. Graham, PhD, Chairman, Department of Food Science and Nutrition, College of Agriculture, University of Missouri, Columbia, Mo.

Arthur W. Hansen, PhD, Del Monte Corp., San Francisco, Calif.

R. Gaurth Hansen, PhD, Academic Vice President, Utah State University, Logan, Utah

William A. Hardwick, PhD, Anheuser-Busch, Inc., St. Louis, Mo.

Robert W. Harkins, PhD, Grocery Manufacturers of America, Inc., Washington, D.C.

Raymond H. Hartigan, PhD, Kraftco Corp., Glenview, Ill.

Warren E. Hartman, PhD, Worthington Foods, Inc., Worthington, Ohio

Richard E. Hein, PhD, General Manager of Research and Development, Heinz, U.S.A., Division of H. J. Heinz Co., Pittsburgh, Pa.

Edward G. High, PhD, Meharry Medical College, Nashville, Tenn.

Roy G. Hlavacek, Food Processing, Chicago, Ill.

Hartley W. Howard, PhD, Director of Technical Services, Borden, Inc., New York, N.Y.

Eugene E. Howe, PhD, Merck and Co., Inc., Rahway, N.J.

Imri J. Hutchings, PhD, General Manager of Research and Development, H.J. Heinz Co., Pittsburgh, Pa.

Ogden C. Johnson, PhD, Acting Director, Office of Nutrition and Consumer Sciences, Food and Drug Administration, Department of Health, Education, and Welfare, Washington, D.C.

Paul E. Johnson, PhD, Executive Secretary, Food Protection Committee, National Academy of Sciences—National Research Council, Washington, D.C.

Albert J. Karas, McCormick & Co., Inc., Hunt Valley, Md.

Oral Lee Kline, PhD, Executive Officer, American Institute of Nutrition, Bethesda, Md.

John A. Korth, Manager, Food Information, Corn Products Co., Englewood Cliffs, N.J.

Carl H. Krieger, PhD, Vice President, Product Research, Campbell Soup Co., Camden, N.J.

Paul A. LaChance, PhD, Department of Food Science, Rutgers University, New Brunswick, N.J.

E. Earl Lockhart, PhD, Assistant to the Vice President, Corporate Technical Division, The Coca-Cola Co., Atlanta, Ga.

Caro E. Luhrs, MD, Medical Advisor to the Secretary, Office of the Secretary, United States Department of Agriculture, Beltsville, Md.

John G. Martland, Vice President, Research and Development, Green Giant Co., Le Sueur, Minn.

Walter Mertz, MD, Chairman, Nutrition Institute, Agricultural Research Service, United States Department of Agriculture, Beltsville, Md.

Walter H. Meyer, The Proctor & Gamble Co., Winton Hill Technical Center, Cincinnati, Ohio.

Lloyd Miller, Carnation Co., Los Angeles, Calif.

Margarita Nagy, Research Associate, Department of Foods and Nutrition, American Medical Association, Chicago, Ill.

Albert S. Neely, Kellogg Co., Battle Creek, Mich.

Robert O. Nesheim, PhD, Vice President, Research and Development, Quaker Oats Co., Chicago, Ill.

Jerry Proctor, PhD, Kraftco Corp., Glenview, Ill.

Dorothy Rathmann, PhD, Corn Products Co., Englewood, N.J.

James M. Reed, Assistant Research Director for Information, National Canners Association, Washington, D.C.

Virgil B. Robinson, DVM, PhD, Dow Chemical Co., Zionsville, Ind.

Dudley Ruch, Vice President for Consumer Research, Pillsbury Co., Minneapolis, Minn.

Irving I. Rusoff, PhD, Nabisco, Park Ridge, Ill.

Harold H. Sandstead, MD, Department of Biochemistry, Vanderbilt University School of Medicine, Nashville, Tenn.

Herbert P. Sarett, PhD, Vice President, Nutritional Science Resources, Mead Johnson Research Center, Evansville, Ind.

Arthur T. Schramm, PhD, Food Materials Corp., Chicago, Ill.

William H. Sebrell, Jr., MD, Director, Institute of Human Nutrition, College of Physicians and Surgeons, Columbia University, New York, N.Y.; and Medical Advisor, Weight Watchers International, Inc., Great Neck, N.Y.

Frederic R. Senti, PhD, Agricultural Research Service, United States Department of Agriculture, Beltsville, Md.

Horrace L. Sipple, PhD, The Nutrition Foundation, Inc., New York, N.Y.

Ira I. Somers, PhD, Executive Vice President, National Canners Association, Washington, D.C.

Elwood W. Speckman, PhD, Director of Nutrition Research, National Dairy Council, Chicago, Ill.

Fredrick J. Stare, MD, Professor of Nutrition and Chairman, Department of Nutrition, Harvard University School of Public Health, Boston, Mass.

Malcolm R. Stephens, Institute of Shortening and Edible Oils, Inc., Washington, D.C.

F. M. Strong, PhD, University of Wisconsin, Madison, Wis.

R. Thiessen, Jr., PhD, Nutrition Research Area Manager, General Foods Corp., White Plains, N.Y.

A. LeRoy Voris, PhD, Executive Secretary, Food and Nutrition Board, National Academy of Sciences—National Research Council, Washington, D.C.

Kenneth G. Weckel, PhD, Professor of Food Science and Industries, Department of Dairy Science, University of Wisconsin, Madison, Wis.

Sidney Weissenberg, United States Federal Trade Commission, Silver Springs, Md.

Philip L. White, ScD, Secretary, American Medical Association's Council of Foods and Nutrition, Chicago, Ill.

Harold L. Wilcke, PhD, Ralston Purina Co., St. Louis, Mo.

Virgil O. Wodicka, PhD, Director, Bureau of Foods, Food and Drug Administration, Department of Health, Education, and Welfare, Washington, D.C.

Stephen A. Ziller, PhD, The Proctor & Gamble Co., Winton Hill Technical Center, Cincinnati, Ohio.

All participants in the Symposium on Vitamins and Minerals in Processed Foods took part in discussions of five task forces on the third and final day (Mar. 24, 1971) of the meeting. Each task force was given a specific charge, to which members were asked to address themselves in preparing recommendations.

Task Force I Needs For Future Research
C.O. Chichester, Chairman

John C. Ayres	Virgil B. Robinson
Myron Brin	Harold H. Sandstead
Gerald F. Combs	Frederic R. Senti
Philip H. Derse	F.M. Strong
Walter Mertz	A. LeRoy Voris

Task Force II Fortification of Processed Foods
Robert O. Nesheim, Chairman

Bernard Alexander	Richard E. Hein
Vigen K. Babayan	E.E. Howe
Evan F. Binkerd	Carl H. Krieger
Ann L. Burroughs	Paul A. LaChance
Norman W. Desrosier	John G. Martland
Richard P. Farrow	William H. Sebrell
William A. Hardwick	Stephen A. Ziller
Warren A. Hartman	

Task Force III Nutrition Education of Consumers
Philip L. White, Chairman

John J. Birdsall	Roy G. Hlavacek
Robert H. Cotton	Dorothy Rathmann
Robert W. Harkins	Horace L. Sipple
Edward G. High	Elwood W. Speckman

Task Force IV Regulatory Aspects of Nutrition Guidelines
Malcolm R. Stephens, Chairman

William A. Brittin	Ogden C. Johnson
J.A. Campbell	Oral Lee Kline
William E. Dickinson	John A. Korth
Donald F. Emery	Walter H. Meyer
Ray C. Frodey	Lloyd Miller
D. M. Graham	Albert S. Neely
Arthur W. Hansen	Arthur T. Schramm
Raymond H. Hartigan	Sidney Weissenberg
Hartley W. Howard	

Task Force V Rationale and Criteria For Micronutritional Improvement of Foods
William J. Darby, Chairman

Ray H. Anderson	Willis A. Gortner
Lawrence Atkin	R. Gaurth Hansen
Richard H. Barnes	Paul E. Johnson
Howard E. Bauman	Irving I. Rusoff
Benjamin Borenstein	Herbert P. Sarett
Gary E. Costley	Harold L. Wilcke
Lloyd J. Filer